MATH BRIDGES TO
A BETTER FUTURE:

MATH BRIDGES TO A BETTER FUTURE:

Using Math to Enhance Family
Communication and Decision Making

James Elander

Rev. date: 05/05/2023

To order additional copies of this book, contact:
Xlibris
844-714-8691
www.Xlibris.com
Orders@Xlibris.com
843971

MATH BRIDGES TO A BETTER FUTURE:
Using Math to Enhance Family
Communication and Decision Making

A book written for adults and families to encourage
cooperation and inter-generational learning, where
adults and children explore math and logic together
using practical, real-world problems.

Suggestions

(Please Read)

This book uses math as a tool for building bridges toward improved family communication and better decision making. It encourages us to be better informed and more contributing members of families and society. The author bridges important mathematical concepts and decision making skills with our family life, since these are critical for everyday living, and are sometimes neglected in this digital age.

The book can be used by families that consist of only adults, as well as by parents with children, and it is appropriate for children of grade school, high school, or college age. It is geared towards adults who want to renew and ramp up their math background, people of all ages who want to bridge their math, logic, and decision making skills to create a life with a better future, and parents who want to help their children improve math skills and learn life lessons. It includes the **How** and **Why**, plus activities called **Investigations** and **Applications** that make this text a valuable and meaningful resource for teachers, also.

Yes, it does involve math, but isn't math involved in almost everything today? For example: banking, taxes, time zones, computers, engineering, shopping, budgets, travel, fitness, sleep quality, measurements, money, competition, and memory.

Bridging Question: What are everyday situations where you use math, logic, numbers, and thinking related to decisions?

The field of Mathematics called Geometry was founded over 2,500 years ago, and it was selected by Socrates and others as the model for decision making. It was taught to prepare future leaders and good citizens, and used ever since for that objective. THINK ABOUT IT! More on this historical development later.

People learn by doing, and the **practical applications in this book** make the material more meaningful to everyday life. The problems selected are intended for people with a general education, not just for mathematicians. The objective is to appreciate and understand the many contributions of mathematics in the various areas of life and, especially, for decision-making. The Bridges (chapters) leading to the applications will broaden a person's knowledge and result in a more informed family member and contributing citizen in society. This is why the term "Bridges" was selected as a way to the future, "ramps" was selected as approaches instead of the word "sections".

To get the most out of this book, the recommended process is to use **small group study and collaboration.** It is suggested to work in small groups (3 to 5 people) and meet for an hour each week over coffee, tea, after dinner, or even for a pre-dinner happy hour. Feel free to keep notes or a journal about your inquiries and discoveries each week, since writing things down may help with understanding and retention. Working in small groups provides a great opportunity for questions and discussion.

No need to work sequentially! You may be selective and work just a few of the problems that interest you, instead of going sequentially through the book. You may choose to have each person take a turn choosing the weekly subject, based on their interests and what problems are most relevant to them. And you may also have each person take a turn leading the group discussions. Your friends may also be interested in joining you, for topics that interest them.

The whole family, husbands, wives, and children, whether working or retired, can be involved in discussions prompted by this book. Children should be encouraged to ask many questions, and perhaps be encouraged to lead the group discussions at times. If you are a family of only one person, this book is for you too! Invite a few friends over each week to work together on questions and discussions.

Encourage questions and explanations as to **WHY** and **HOW** rather than just memorize material. It may even be advisable to review a concept or method with exercises at another time after it is first studied. This depends on how important the concept is to you. Example: You may think the concept of odds, probability, and fair bet are more important if you live in an area where there are many casinos.

It is suggested that one topic per week be covered. Consider starting with something simple like interesting family history, to get the process rolling. Don't try to tackle too much at once, remember, the small group collaborative process should be an enjoyable learning tool. For example, to start slowly, you might decide to have each person review a short history of their lives, for the first few weeks. Future topics could be

---

bank accounts and investments, insurance, health care, car ownership and licenses, utilities, money, anything from everyday life. A different topic can be investigated and discussed every session. They can be related to children's school work or adult's life events. The timing and length of time for discussion is flexible and is suggested not more than one night per week, and about an hour at a time.

So, in addition to the Bridges and Ramps in this book, these weekly group inquiries can branch out to include any other life topics as well, such as:

- family history
- when you were in school? (how it differed?)
- where you grew up
- at was your first job?
- the first date, what made it interesting and meaningful?
- investments
- house buying
- homeowner's insurance
- favorite vacations
- health care
- Long term care programs (what is not included in some plans)
- retirement plans
- when to sell a house
- mortgages and reverse mortgages
- car buying and car insurance
- banking (look up the rule of 72).

For some of the topics such as health insurance, life insurance, and financial investments you might want to invite an agent, or broker, to come in for answers and explanations.

The author suggests not including any "tests" at the end of each session, as tests can create problems and pressure, and usually do not enhance understanding.

Mature groups tend to learn better without tests, since explanations and discussions will be much more valuable. Each person should write their summaries and take notes. Some of the author's most interesting evening college classes consisted of students from age 15 to 50. So, the more diversity in ages and perspectives, the better!

Getting Started

For each Bridge you cross, it is recommended that you always read the material labeled **Basic**. Read it at least once, since it contains information and explanations that will be beneficial to you as a "student-professor" in your group and family.

After reading these Suggestions, you are now ready to review the Table of Bridges, to get a visual picture of the topics covered in this book. You will find most Bridges (chapters) are very short, except a few Bridges at the beginning, since they are a review. This "Bridge" structure is also divided into smaller understandable sections, labeled Ramps.

Enjoy taking these Ramps, up to the Bridges, which will hopefully lead you to improved decision making and a better future!

Suggestions for other material to read.

The Bibliography lists many books which will surprise you (such as **Flatland** written in the 1890s, and still an interesting read today). This is why the Bibliography is at the beginning, so the readers may be aware of these books and possibly read and enjoy this suggested material.

The Author

(A lot older now, but a better golfer.)

James Elander (retired), former Assistant Professor of Mathematics at North Central College, Naperville, Illinois.

Prior to college teaching, he was the Mathematics Department chair at Oak Park and River Forest High School, Oak Park, IL. His credentials include:

- ❖ Western Illinois University,
- ❖ Illinois State University,
- ❖ General Electric Fellow at Purdue University, one summer,
- ❖ Western Michigan University (National Science Foundation Programs),
- ❖ Burlington Northern Achievement Award (North Central College),

- ❖ Past President of the Metropolitan Mathematics Club (Chicago),
- ❖ Past President of the School Science And Mathematics Association (SSMA),
- ❖ Past Chairman of the Illinois Section of the MAA Geometry Committee,
- ❖ Served on many North Central Association evaluation teams.

He is the author of GEOMETRY FOR DECISION MAKING, 1992.

He has also authored four additional books:

Book 1: TGIF-MATH (Also in paper back and e-book at your local book stores.) A text consisting of over 200 activities designed to make class days prior to days like Homecoming, more enjoyable and yet a learning situation. Adults have found these activities very intriguing and enjoyable as family activities.

Book 2: BASIC HIGH SCHOOL MATH REVIEW with Decision-Making A review of Algebra and Geometry in preparation for college. Available in paper back. Designed for students who took only Algebra and Geometry and now feel they need an additional course.

Book 3: TEAM WORKING FOR BETTER MATH STUDENTS: A COLLABORATIVE APPROACH FOR PARENTS, TEACHERS, AND STUDENTS IN GRADES 1-12 (This Text is also written for Home School programs.)

This is a book for grades 1 to 12 yet this book will help the student more in the first 7 or 8 grades, and, I would predict the parents will need help from the students in grades 9 to 12. All the books cover the needed life time skills of Critical Thinking and Decision Making.

Book 4: Plane and Solid Geometry Essentials A new text covering the essentials of PLANE and SOLID GEOMETRY with Decision Making. This is a new text, published in 2022. It covers the essentials of Plane and Solid Geometry in one year. The author found it helpful to teach geometry in one year, so calculus could have the whole senior year.

Book 5: A Book For Thinkers TO IMPROVE FAMILY RELATIONS WITH DECISION MAKING METHODS IN A CHANGING WORLD!

Rodin's THE THINKER outside the Legion of Honor, San Francisco, CA.

Prologue

This text is written as a result of observing and working with students at the high school and college levels for over 35 years. Some of the college classes had an age range from 15 to 60. This book is designed for people of all ages who are interested in the mathematics that is needed or required for areas of business, accounting, economics, psychology, and the trades. It also provides a general education in mathematics for intelligent citizenship with basic decision making skills. Whatever you do, you will have to think to be successful. The author's objective is to provide an introduction and/or review of these basic mathematical needs in a readable manner with realistic explanations as to the why and how. Such questions are: Why you need to keep up to date and review math occasionally? What are some of these needs? Added are interesting bits of Math History such as why **Pi** is equal to 3.14159...? What is the history behind Pi? Who named the constant **Pi and why**?

Bits of History of these related topics makes the book very interesting. Another example: What did the State of Indiana try to do in the 1800s related to pi? This interesting bit of history will be discussed later, and you may be surprised by some titles in the Bibliography such as, MITS, WITS AND LOGIC and THE EDUCATION OF T.C. MITS.

People enjoy learning new material when they understand it. Understanding is like an optical illusion, now you see it, now you don't! Since mathematics builds on past understandings, the first Bridge is a review to provide and ensure the basics for the new material. Parts of Bridge 1 may be skipped or rapidly covered if you feel you know the material. This was intentional and carefully done to provide some drill, some challenges, and some forgotten material, which may also be new. Another Bridge is the basic mathematics of finance. The reason for this is simply the fact that people are interested in dealing with money matters, like the Bankers Rule of 72.

A unique feature of the text is the optional **investigations** at the end of most Ramps. These investigations review topics from areas or branches of geometry via **applications**, and more importantly challenge the reader to **think**. This text provides the opportunity for you to select applications you

---

recognize and perhaps have encountered and may be more interested. (Many of these unique problems will be interesting to you and your family.) There are many problems and you will have to be selective.

The multitude of comments add to the understanding and appreciation by referring to math history and/ or providing new ideas for investigations. These are valuable for all members of the family. The Pi problem mentioned above took over 4000 years to finally complete!

The key to understanding is the fact that the text is mathematical (or *decision making*) in its development. This means the approach is from undefined terms, defined terms, postulates, to theorems, with applications for better understanding. Many ANSWERS are given either following a problem or at the end of the activity (some answers may even be wrong!), and you may need to explain your answer or your justification. This will provide immediate feedback for better understanding. The "correct answer" by your justification can be gratifying.

Keeping a notebook, or marking the book, is encouraged but tests are not! Learning and understanding are improved by questions and discussions. Many people first question the idea that there are no tests, but if they are asked to recall the formula to calculate a mortgage payment or the volume of a sphere, they will change their viewpoint. (Employers would prefer their employees to use their notes and arrive at the correct answer, rather than submit the wrong answer.)

There are more exercises than you have time to work, so be selective, pick problems that are related to your interests. ***The function of this book is not*** *math problem solving, but to improve your critical thinking ability.* (An effort was made to select applied or practical problems for a variety of backgrounds.) The text assumes a background of one year of Algebra and Geometry. The following tools will be useful:

- Pencils,
- Ruler and Protractor,
- Graphing paper,
- Scientific calculator,
- Lined paper,
- and a Notebook.

The author has attempted to assist you by the method **"We learn the new in the light of the old"** and to answer the question, **"Isn't there an easier way to work the problem?"**. Completing summaries, seeking help, and understanding your errors is logical and gratifying. When in doubt read the material again, ask for help from your group and friends, possibly even your children, and you will succeed! You will find the course very informative, useful and, to varying degrees, even enjoyable. Hopefully, you will have a small group so there will be discussions. Remember,

"Mathematics is not a spectator sport."

Take a glimpse at the Table of Bridges after reading this prologue. A unique bibliography of selected titles is provided and includes suggestions for additional and very interesting readings. Hopefully you will take some time to explain the readings to others.

If you have a question, or a problem arriving at an answer, ask a member of your group to discuss their solution. (Answers are given for many problems and some answers are intentionally wrong.) The answers may differ, due to entering rounded data, but your thinking could be correct. The ideal group is about 5 or 6 people, which would meet once a week to ask questions and discuss the concepts plus do the problems. You may have to justify your thinking to the group. Reminder, there are many more problems listed then you need to work. So, once you understand the method, look for the problems that pertain to your interests.

Good luck in your "bridge" building!

The author appreciates the many colleagues and students who, unknowingly, advised and challenged the author to search for explanations and better ways, and also notes the many benefits and ideas from attending professional meetings. He also enjoyed getting input from Olaf and Jaime, as that was very helpful!

The next section or Bridge is a list of very useful and enjoyable books. Notice the first one, **Flatland** by Abbot, written around 1890. I would suggest this be your first group discussion after one person explains the book. Perhaps a young woman might read it, comment on the book and answer the questions. All will be surprised!

Selected Bibliography

*FOR STUDENTS, PARENTS AND TEACHERS,
ENYOYABLE and INFORMATIVE READING*

**Comment: Look at some of the interesting titles,
like FLATLAND by ABBOTT**

Aichele, D. B.
  "Using Construction Tools to Discover Mathematics"
  School Science and Mathematics,1975

Banks, Robert B.
  **SLICING PIZZAS, RACING TURTLES, AND FURTHER
  ADVENTURES IN APPLIED MATHEMATICS,** Princeton
  University Press, 1999

Beckmann, P.
  HISTORY OF PI, Golden Press, 1971

Bell, E. T.
  MEN OF MATHEMATICS, Simon & Schuster, 1937

Bergamini, D
  MATHEMATICS, Time, Inc. Book Division 1963

Byrkit, D.
  "TAXICAB GEOMETRY" MATHEMATICS TEACHER, May 1971,
  Pages 418 -422.

Dantig, T.
  THE BEQUEST OF THE GREEKS, Charles Scribner's
  Sons, 1955

Davis, P. and Hersh, R.
  THE MATHEMATICAL EXPERIENCE, Houghton Mifflin,
  1981

Dunham, William
   JOURNEY THROUGH GENIUS, John Wiley and Sons, 1990.

Dudley, Underwood
   NUMEROLOGY or What Pythagoras Wrought, Mathematical Association of America, 1997

Dudley, Underwood
   MATHEMATICAL CRANKS, Mathematical Association of America,1992

Eves, H.
   GREAT MOMENTS IN MATHEMATICS BEFORE 1650, Mathematical Association of America,1983

Fadiman, Clifton
   THE MATHEMATICAL MAGPIE, (Mobius Strip-Paul Bunyan vs. The Conveyor Belt), Simon and Schuster, 1981

Fawcett, Harold
   NATURE OF PROOF, 13th Yearbook of NCTM

Gardner, M
   MATHEMATICAL CARNIVAL, Alfred A. Knopf, 1975

Gazale, Midhat
   NUMBER: From Ahmes to Cantor, Princeton University Press, 2000

Gordon, Sheldon and Florence, Editors
   STATISTICS FOR THE TWENTY-FIRST CENTURY, Mathematical Association of America,1992

Gross H. and Miller, F.
   MATHEMATICS A CHRONICLE OF HUMAN ENDEAVOR, Holt, Rinehart and Winston, 1971

Hahn, L.
COMPLEX NUMBERS AND GEOMETRY, Mathematical
Association of America,1996

Hallerberg, A.
"The Metric System: past, present - future?",
Arithmetic Teacher (April 1973) 247–255

Huff, Darrell
HOW TO LIE WITH STATISTICS, Norton, 1954

Johnson, D.
"Paper Folding for the Mathematics Class" NCTM,
1957

Kasner and Newman
MATHEMATICS AND THE IMAGINATION, Simon & Schuster,
1940

Kenner, Morton R
"Hemholtz And The Nature Of Geometric Axioms",
Mathematics Teacher, Vol. 50, Feb. 1957

Klein H. A.
THE WORLD OF MEASUREMENTS
Publisher: Simon a Schuster, 1974

Kline, M.
MATHEMATICS AND THE PHYSICAL WORLD
Thomas Y. Crowell Co., 1959
Comment: Chapter 7 Trigonometry

Kline, M.
MATHEMATICS IN WESTERN CULTURE
Oxford University Press, 1953

Konhauser, D., Velleman, D., Wagon, S.
WHICH WAY DID THE BICYCLE GO?

Mathematical Association of America, 1996

Kramer E.
THE NATURE AND GROWTH OF MODERN MATHEMATICS
Hawthorn Books, Inc. 1970

Lieber, L.
MITS, WITS, AND LOGIC
Institute Press, New York, 1954

Lieber, L.
THE EDUCATION OF T. C. MITS
W. W. Norton & Co., 1954

Loomis, E.
THE PYTHAGOREAN PROPOSITION
NCTM publication

Maor, Eli
THE STORY OF NUMBER
Princeton University Press, 1998

Maor, Eli
TRIGONOMETRIC DELIGHTS
Princeton University Press

McCamman, C.
"Curve Stitching in Geometry"
NCTM Yearbook: **MULTI-SENSORY AIDS IN TEACHING OF MATHEMATICS**

Meyer, J.
FUN WITH MATHEMATICS
Chapter: **"Curves That Control Our Lives"**, Dover Publications, 1952

Miller, C. D.
MATHEMATICAL IDEAS

Scott, Foresman & Co., 1986

Niven, Ivan
THE MATHEMATICS OF CHOICE
Mathematical Association of America, 1965
Nolan, Deborah, Editor
WOMEN IN MATHEMTICS: SCALING THE HEIGHTS,
Mathematical Association of America,1997

Olson, A.
MATHEMATICS THROUGH PAPER FOLDING
National Council Teachers of Mathematics, 1975

Packel, Edward
THE MATHAMATICS OF GAMES AND GAMBLING
Mathematical Association of America, 1981

Parker, Maria
SHE DOES MATH!
Mathematical Association of America, 1995

Poe, Edgar Allen
THE GOLD BUG
(A Mystery involving mathematical reasoning.

Pohl, V.
HOW TO ENRICH GEOMETRY USING STRING DESIGNS
NCTM, 1968

Polya, G.
MATHEMATICAL DISCOVERY, Vol 2
John Wiley & Sons, 1965
Chapter 14: The art of teaching Mathematics

Postman, N.
TECHNOPOLY (Very interesting-adults)
Alfred A. Knopf, 1992

———
23

Reid, Constance
   JULIA, A life in Mathematics
   Mathematical Association of America, 1996

Reid, Constance
   FROM ZERO TO INFINITY
   Mathematical Association of America, 1992

Reid, Constance
   A LONG WAY FROM EUCLID
   Thomas Y. Crowell Co., 1963

Schell, Michael J.
   BASEBALL'S ALL-TIME BEST HITTERS
   Princeton University Press, 1999

Schmalz, Rosemary
   OUT OF THE MOUTHS OF MATHEMATICANS
   Mathematical Association of America, 1993

Simon, W.
   MATHEMATICAL MAGIC
   Charles Scribner & Sons, 1964.
   (Magic squares and a method for solving them.)

Steinhaus, H.
   MATHEMATICAL SNAPSHOTS
   Oxford University Press, 1950

Sterrett, Andrew, Editor
   101 CAREERS IN MATHEMATICS
   Mathematical Association of America, 1996

Stevenson, R. L.
   TREASURE ISLAND (chapter 31)
   (Locus problem-location of the treasure.)

Taylor, Alan D. and Zwicker, William S.
   SIMPLE GAMES, Princeton Press, 1999

Turnbull, H. W.
   THE GREAT MATHEMATICIANS
   New York University Press

Weber, R.
   A RANDOM WALK IN SCIENCE
   Crane, Russak & Co. Inc., 1973
   (Fascinating little story on p. 124 with a surprise
   ending. "Life on Earth by a Martian")

Videos

DONALD IN MATHMAGIC LAND
   Disney

MATHEMATICALLY MOTIVATED DESIGNS
   Eves, Howard Mathematical Association of America

CRITICAL THINKING TEST, LEVEL X
(A very unique test)
   R. Ennis and J. Millmani:
   www.criticalthinking.org.
   Critical Thinking Test and Software
   1-800-833-364 or 1-800-458-4849

Table of Bridges and Ramps

Bridge 1: Math — Some Old/Some New

"MATHEMATICS IS THE GATE AND THE KEY TO ALL SCIENCES. HE WHO IS IGNORANT OF IT CANNOT KNOW THE MATH THINGS OF THIS WORLD"

Roger Bacon

The recommended process for this book is to use **small group study and collaboration.** It is suggested to work in small groups (3 to 5 people) and meet for an hour each week over coffee, tea, after dinner, or even for a pre-dinner happy hour. Perhaps your children can invite some of their friends to join you.

There are more problems than you need or would want to try to work, so select the ones that are interesting to you.

Comment: The author suggests you keep notes, or mark the book, to remind you of important items from the activities for future identification and use. You might indicate the page numbers in your notes so you can easily locate them. These are: definitions, postulates and theorems.

This Bridge will provide you with some reasons why people should periodically review and update their Math. In this chapter we cover some basic skills, review four types of numbers, review their properties and operations, plus update business and industrial applications. The history of MATHEMATICS will review the main reason for teaching Geometry, which is to teach Decision Making. (This main reason was created by philosophers like Socrates about 5 hundred BC or BCE.)

At the end of Ramp 1 there is a 1936 Civil Service Exam. Your parents or friends, or even grandparents, may be interested in looking at it. You may want to read over the test and you will probably be surprised as to some of the problems. Future teachers had to spend a week taking exams in order to qualify as a teacher in the early 1900s. The math exam could take all day.

Ramp 1.1
Why Use This Book?

Comment: You may wish to skip all or parts of Bridge 1 since much of it is a review.

On the other hand, it is suggested you read each Bridge in order to have a better idea as to the author's objectives. Completing this bridge will provide you with a better understanding as to what you learned and perhaps have forgotten over the years. It is suggested that you read all the material, even the activities, since many problems are thinking and informing ones. Some problems you may find very different, but interesting. You don't have to do all the problems.

The first two Ramps (sections) of this Bridge are a review of previously studied concepts and skills requiring not only recall but also explanations as to why. Some unusual thinking type problems and a few other investigations will be explained. The reason for this comment is to caution you against a possible first impression that this text is too elementary or too difficult. Take a look at the headings in the Table of Bridges for a quick overview of topics, and then make the decision. Are you able to justify your decisions? Can you logically present and defend your decisions or position on issues? The author believes all people can think and make valid logical decisions.

Question to ponder: What makes a decision valid?

You will find the new topics in this text not only beneficial, but also very useful in understanding how practical mathematics is able to assist you in making non-mathematical decisions.

Mathematics is a bridge to your future and is basic for decision making. You are probably thinking this is an odd way to begin the study of a mathematics related course. The author's aim is to create some curiosity, so you will start this study with a positive attitude. The attitude you have about a course will, to a large degree, determines your success. Attitude does make a difference! You must remember grades are nothing more than a measure or indicator with regard to past performances. It may have nothing to do with your future accomplishments. (For example, Ray Kroc dropped out of high school yet started a hugely successful restaurant. Who was Ray Kroc?) Your success in these bridges is not determined by performance in your last math program, but only by what you do or learn in this one.

Comment: Ray Kroc, founder of the McDonald food chain, grew up in Oak Park, Illinois. This could be a topic for a report.

The former chairman of the World Trade Corporation advises youth who are hopeful for a bright future to study mathematics for its bi-products.

"Young people who have acquired the ability to analyze problems, gather information, put the pieces

together to form tentative solutions will always
be in demand."

J. G. Maisonrouge
Board Chairman
IBM World Trade Corp.

Activity 1.1

Answer the following questions.

1. From the above quote, what would you say young
   people should know how to do?
2. What were the objectives of your high school
   geometry course? (From your memory?)
3. The following is The Preamble of United States
   Constitution:

"We the people of the United States, in order
to form a more perfect Union, establish justice,
ensure domestic tranquility, provide for common
defense, promote the general welfare, and secure the
blessings of liberty to ourselves and our posterity,
do ordain and establish this Constitution for the
United States of America."

   a. List from the above the terms or words that
      you classify as undefinable. How many are
      there?
   b. List the number of words that you are
      classifying as definable. How many?
   c. How many words are there in The Preamble
      quote above?
   d. What percent of the total number of words
      are undefinable?

4. Simplify one of the following as to the number
   it represents:

a. 6+8x4-2(3-5) + 60-10

b. 36-12x3+27/3+6

Answers: a. 92 b. 15

5. Ask a few of your friends to work problem 4. Then compare your answers with theirs. Are the answers identical?
6. Work problem 4 using a calculator.

a. Are the answers the same as when you calculated manually?
b. If they are not the same, then attempt to explain why.
c. The answers should be the same. Be sure to explain the reason for the difference in the answers, if necessary, recall the Order of Operations.

Ramp 1.2
A Bit of History

The order for performing the operations (x and ÷, then + and -) to simplify operations was established approximately 3000 years ago 1000 BCE in ancient Egypt by the bankers.

To understand the why for the order of operations, physically proceed to calculate the value of a quantity of coins in your piggy bank. What do you do first, second, and so forth? This will illustrate why we multiply and then add to evaluate a set of arithmetic operations.

Suggestion: Do the above with your friends. (This is a good activity to justify the order of operations when you try to determine the value of a number of coins in a piggy bank.) Many students were taught the order of operations by the first letters in the following statement: "**M**y **D**ear **A**unt **S**ally." (x,÷,+,-)

Historical Note

About 1000 B.C. the bankers of Egypt informed the business world and the mathematicians employed by the banks that they had to use the **MDAS** method for exchanging monetary values. The banks in Egypt were the center for the businesses from Europe and China. (A very interesting non-fiction book related to this is *1421: THE YEAR CHINA DISCOVERED AMERICA.* The author is Gavin Menzies. Not a math book, but related to the history of China.)

Thinking Problems

1. Each letter in the following problem represents a unique digit. Replace the letters with digits (0-9) so that the addition problem will be correct. If T is 5, then no other letter can be 5. What digit must R be in the following?

$$H \; I \; T \; + \; H \; I \; T \; = \; R \; U \; N \; S$$

Comments:
   a. Yes, there is more than one answer.
   b. But what must be the value for R? Why?
   c. What is the one possible answer?
   d. These are fun thinking exercises.

2. The following data is the result of interviewing a group of parents:

   ❖ 16 were driving Ford Products,
   ❖ 21 were driving General Motors products,
   ❖ 15 were driving Chrysler products,

❖ 4 were driving Fords and General Motors products,

❖ 3 were driving General Motors and Chryslers,

❖ 5 were driving Fords and Chryslers,

❖ 2 were driving all three products.

a. Draw the **Venn diagram** depicting the above information. (Three separate circles that all intersect so there are 7 sections. Place the above information in the proper section.)

b. How many were driving only Fords? Only GMs? Only Chryslers?

Answers: b. Only Fords 9, Only GMs 16, Only Chryslers?

Fill in the area! Example: The 2 is filled in representing only 2 drove all three makes.

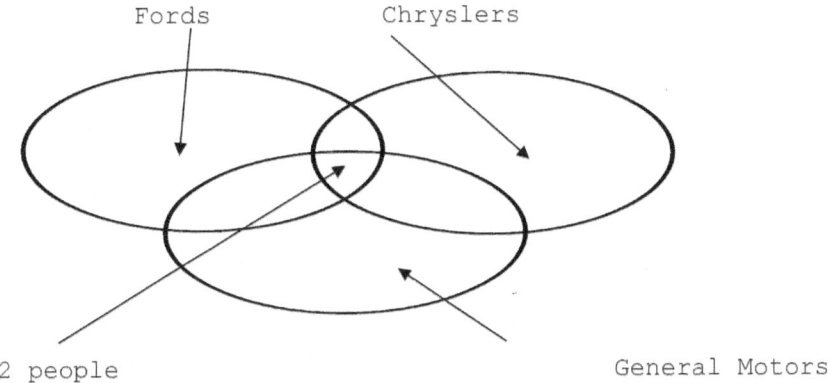

Fords   Chryslers

2 people   General Motors

Number Problems

3. Why is the answer to a negative number times a negative number a positive number?

Comment: **It is a necessary conclusion that the answer is positive!**

The following is one explanation to illustrate why.

We know 4x5 = 20. Now try it this way!

(5-1) x (6-1)
we know the answer is 20.
   (5-1)x(6-1) =
   5(6)+(5)(-1)+ (-1)6 +(-1)(-1)=
   30 +(-5)+(-6) + (-1)(-1) = 20

Simplifying:
   30+(-5)+(-6)+(-1)(-1)=
   30 - 11 + ? = 20 =
   (-1)(-1)must be 1.

................this requires -1 x -1 to be equal to +1!

Explain this to another person!

4. Question: Why can't you divide by zero?

   Comment: Let N be a non-Zero number so that N/0 = A, where A is the answer. Check the answer:

   N = 0xA. The product will always be zero instead of N: therefore, this illustrates division by 0 is impossible.

5. Classify the following Geometric terms as always true or false. Be able to explain your answer. **(Class discussion!)**

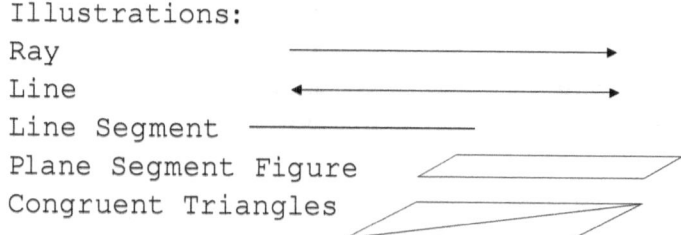

    Illustrations:
    Ray
    Line
    Line Segment
    Plane Segment Figure
    Congruent Triangles

  a. Is a ray and a line the same?

b. What is a plane?

c. What is a point?

d. What is a Plane segment figure?

e. True or False: Congruent triangles are always similar.

f. True or False: Figures with equal area are always congruent.

g. True or False: Three points will always determine a geometric plane.

h. True or False: Lines that don't intersect are always parallel.

i. True or False: Three points always determine a triangle.

j. True or False: A 3-4-5 right triangle can be used to illustrate the Pythagorean Theorem.

k. True or False:3 lines can be perpendicular to each other at the point of intersection.

(Discuss the answers and the whys. Drawings or models will help in some cases.)

Hint j: Regarding the Pythagorean Theorem (2500 BC), the mathematicians of Babylon knew the 3-4-5 case but had no proof for the theorem. They also had a value for what is now the Pi constant but could not prove it. The Egyptian surveyors were called the rope Stretchers. Why? (3,4,5)

Hint k: Look at a corner of your ceiling.

Another bit from history: People often ask the question relative to the symbol V used to represent the number 5. This came about from the following: Close your hand and to indicate counting 1 to 5, for 1 you opened one finger, for 2 the second finger, and so forth to 5. Look at your hand for 5. Do you see the V? Now put the two Vs together and you have 10(X). Also, X is on the keyboard.

___

Ramp 1.3
Geometric Review

Answers are at the end of investigations.

1. Draw a 3D picture of a pyramid with a square base. (Use dotted line segments for the hidden segments.)
2. Draw a picture of a pyramid with a square base, flattened out or what is called the layout form or view.
3. The following figure, ABCD, is a square with Diagonals AD and BC intersecting at O.

Some communities arrange 4 softball diamonds in this design to save space and yet provide maximum playing space. The home plate for each diamond is located near O, and the diagonals are fences. Home plates, one at each O.

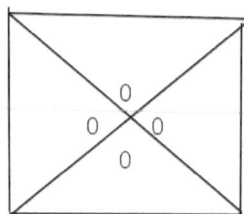

    a. Where do you think benches are for the players, and where for the parents?
    b. Add your details to a much larger drawing, if interested.

Application Thinking

1. A student, who was at summer camp, was given the following problem. He was given the morning chore to go to the pump and bring back four gallons of water to mix with orange concentrate.

But he was given only an 8-gallon, 5-gallon and 3-gallon container. (There are no markings on the containers.)

How did he do it, that is, bring back four gallons of water?

Try to solve for and show the method to end up with 4 gallons.

One answer: Fill the 8-gallon container to start with and then adjust as indicated below, moving water in the following nine steps (some students may find a different solution with a different number of steps):

8 gal	5 gal	3 gal
8	0	0
5	0	3
5	3	0
3	2	3
6	2	0
6	0	2
1	5	2
1	4	3
4	4	0

2. In the following drawing of a pyramid, which segments should be dotted (meaning hidden from view) in a 3-dimensional figure?

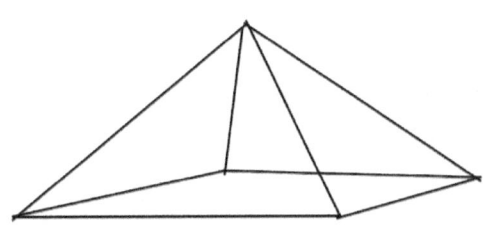

The total surface area can be calculated from the lay out of the pyramid. Area and volume formulas will be justified later, which you will find interesting.

Write your summary!

A point is an undefined term (Points are usually labeled with capital letters.)

A moving point generates a straight line.

A ray is a point moving in one direction.

A line segment is between two points.

All Conclusions consist of undefined terms, defined terms, postulates, and possible other conclusions.

A MATH BOOK develops its conclusions in the same method, using undefined terms, defined terms, postulates, followed by theorems derived from the previous three.

This book is not a text book with rigorous proofs, but in a few cases, it will follow a logical development to help develop your understanding.

In Geometry, one of the first postulates is:

The shortest distance between 2 points is a straight line.

(Label it Postulate 1 if you wish to identify it later on, or copy a note on a blank page at end of the book for a future quick find.)

For example, we can look at this theorem: in triangle ABC, AB plus BC is greater than AC, and also AB is greater than CB – BC, and BC > AC– AB. This can be called a **theorem** since it results from postulates.

Another postulate is that **3 non-collinear points determine a plane.** To better understand this, try to support a plate by one point or your finger, then 2 points (2 fingers) and finally by 3 non-collinear points. Hence, the definition that 3 points can determine a plane will be labeled **Postulate 2.**

What is a Theorem?

In the appendix are listed the Definitions, Postulates, and the Theorems.

All logical systems are based on undefined terms, defined terms, postulates, conclusions that follow, and theorems that follow logically. In government, theorems could be thought of as laws.

Ramp 1.4
Why Study Mathematics

"That they (all citizens) might excel in public discussions on philosophic or scientific questions, or pass laws the people must be educated in philosophy, mathematics, science, and local knowledge."

The above is from Athenian Sophist
School Curriculum (480 B.C.E.)
F. Cajorie

Comment: This introduction is lengthy, but the message is important, so read it carefully and discuss the issues with others in your group or friends.

Students often ask: Why do I need this math course or other courses? This is a valid question. The answer is, all too often, just do it and some day you will understand why. But most of the time it takes an unusual and truly motivated person to study for the pure quest of knowledge.

"Mathematics has been an activity for thousands of years. To some small extent, everybody is a mathematician and does mathematics unconsciously."

Phillip Davis & Rueben Hersh

If students can be convinced of the value of mathematics from the beginning, then they will see that this course will be practical, useful, applicable, and will have the greatest opportunities for employment. All courses require skills involving various areas of mathematics and in different amounts. The following professions all use math, and reviewing this list may lead to some interesting discussion.

Health Services
Primary-care workers
Nutrition counselors
Gerontology specialists
Hotel Management and Recreation

Restaurant workers	Aviation
Resort employees	Waste management
Travel agents	Computers
Conference planners	Maintenance
Food Services	Design
Managers and chefs	Retail sales
Food processing	Consulting services
Food laboratories	Business Services
Engineering	Accounting
Robotics	Statistical analysis

Payroll	Teaching
Word processing	Day-care
Human Services	Mathematics
Personnel relations	Computers
Job evaluation	Foreign language
Benefit planning	Home school programs
Legal assistance	Maintenance and Repair
Public relations	Industrial
Financial Services	Home Business
Insurance	Primary & Secondary
Financial planning	College Science
Banking	

The study of mathematics will give you some of the tools needed for the above along with your other courses involving the skills of: Writing, Drawing, Communicating, Key boarding, Computer literacy, Decision-Making, Team participation, and getting along with others.

If you knew what you would be doing in a few years, your education could better prepare you for it. The fact is you don't know what opportunities will be available, but hopefully you will be prepared. The observation indicates that a person may change jobs every five to ten years on the average and the time cycle is lessening each year. The following quote from a previous section, needs to be repeated to emphasize its importance.

"Young people who have acquired the ability to analyze problems, gather information, and put the pieces together to form tentative solutions will always be in demand."

J. G. Maisonroug,
Chairman of the Board,
IBM World Trade

You may not in the future use all of the topics included in this text, but it can be guaranteed that you will use the majority of them. Perhaps the most important topic is Decision Making, which everybody will use. It was stated above that statements of conclusion are based on or consist of undefined terms, defined terms, postulates and conclusions that follow.

Now the interesting part is that in the last 500 years of BC or BCE, the city of Athens (and I think Sparta also) had a representative form of government. **Socrates and others realized that this form of government required an informed public.** (The government even had a rule that only landowners could vote, no matter race or gender.) These philosophers, Socrates, Plato, Thales, Euclid and many others, developed programs that taught decision-making and to make it short, it was called geometry. Geometry is a subject that involves undefined terms, defined terms, assumptions, and rules (Theorems) that follow logically.

This is why Geometry has been and still is taught and required in the schools! Prof. Harold Fawcett justified that Geometry will teach decision-making if it is taught for that objective. (See the 13th yearbook of NCTM Nature of Proof.) You will be surprised by a 2015 program created by 4 former United States Presidents to improve the country. (This is described later.)

Geometry related Questions
(Remember, Be selective!)

Comment: Ask for help or hints when needed!

1. Draw the 3D figure for each of these layouts. (Use a ruler)

a.

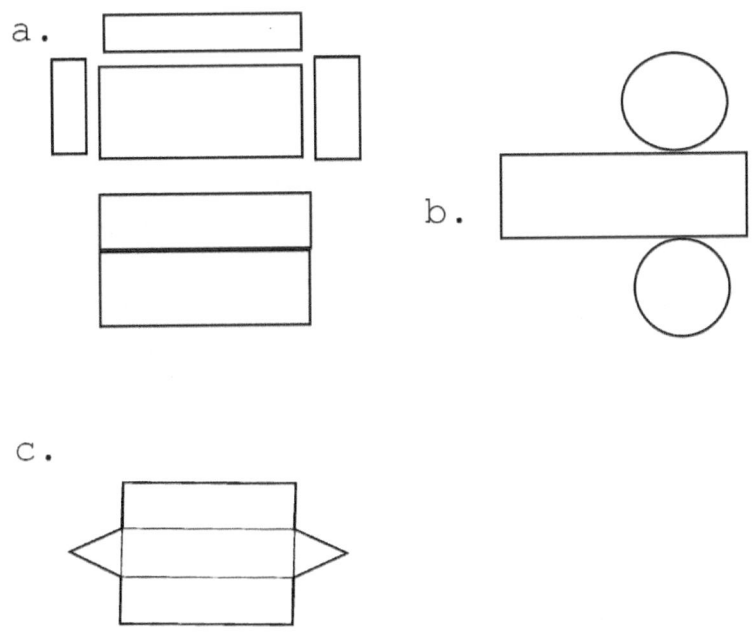

b.

c.

Answers: a. Open shoebox, b. A layout for a can or cylinder, c. Triangular prism

2. What is the measure of side AC, to the nearest foot, in the following drawing of a city park? Hint: ABC is a right triangle. Identify the theorem.

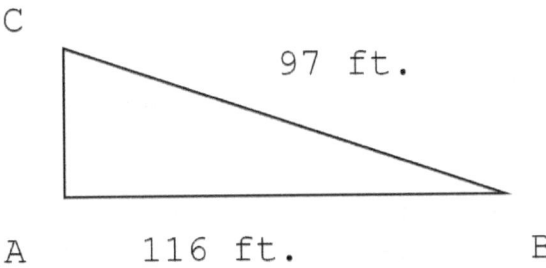

C

97 ft.

A   116 ft.   B

What is the name of this important theorem which will be justified later on with comments?
Name: Pythagorean Theorem

Answers: 5, 50, 500, 5000
L = N(.5)ft. Miles = N(.5)/5280

Link to the conversion tables. See Appendix 4

3. The following statistics are from the Nov. 12, 1991 Chicago Tribune. Workers in the indicated type of work in 1980 and 1990.

Type	Year	
(in Millions)	**1980**	*1990*
Manufacturing	20	225.0
Construction	4.3	5.2
Mining	1.0	0.7
Services	64.7	85.3
Retail trade	15.0	19.7
Government	16.2	18.2
Financial	5.1	6.8
Transportation	5.1	5.8

a. Which type had the largest increase?
b. What is the increase?
c. Did any have a negative increase? If yes, name the type and what was the decrease in thousands of workers?
d. What is the total increase in workers from 1980 to 1990?
e. Which type do you feel involves the least amount of training? Explain your answer.
f. Construct a graph, any type, showing the increase for each type of work. Assume the graph will be printed in the newspaper. Look at some graphs in the newspaper for ideas and types.

Answers:
   a. Services
   b. 20.6 million
   c. yes, mining
   d. 35.1
   e. Services is the most likely answer, but student's explanations are more important than the classification.
   f. Parents check the graph. (Use the computer, if available, also graph for a picture.)

4. What is wrong with the following?

   a. a. If 1 foot is 12 inches or 1 ft. = 12″ and a yard is 36″, we can write: 1 ft. = 1/3 yd. or 12 in. = 1/3 yds. Now multiplying 12 in. by 1/3 yds gives 4 in.= 1/9 yd and taking the square root of each side gives 2 inches = 1/3 yard
   b. Explain the error.

   Answer: The numbers are not identical. An equation is a statement that says two numbers are equal and then the postulates for equations are valid.

5. Why does a 4-legged chair sometimes wobble?

   Hint: Three geometrical planes can be determined by 4 points? List them using A, B, C and D for the points. (4 planes could be formed.)

6. Solve:(give answer to the nearest 2 decimal places):

   $$15 - 5(6x - 7) = 50x + 107$$
   Answer: −.71

Comment: -.71 does not equal -.7125. The reason is the rounded answer was used to check and not the correct answer. Rounding off methods to determine answers is a very important decision. (More on this topic later.)

Why is the price of gas stated as $2.599, instead of as $2.59 per gallon?

7. What is the size of AC? ?> AC <?

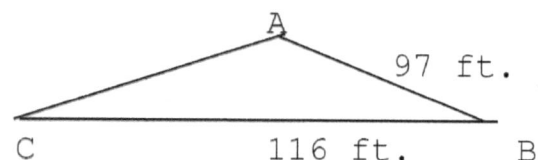

Comment: If the sum of two sides of a triangle is greater than the third side. What do you know about the length of AC? -- < AC <--?

8. Which is a better buy for a dinner, at a steakhouse? Why?

    a. A 10 oz. steak for $25.40.
    b. A 7 oz. steak for $17.70.
    c. A half-pound of steak for $20.32.

    Answer: All three are the same value ($2.54 per oz) so it depends on your appetite.

9. When exceptional service is provided by a waiter, a suggested tip at one time was 10% of the bill. (The word tip is an acronym for "To insure promptness.") What is an easy way to calculate a 10% tip for the following amounts?

a. $7.50   b. 15.50   c. $32.75
d.$42.85   e. $51.24

Answer: Move decimal point 1 place to the left. Explain why.

The trend in recent years has been to leave a 15% tip for exceptional service. What word needs defining? Is there an easy way to calculate a 15% tip for the following amounts?

a. $7.50   b. 15.50   c. $32.75
d. $42.85   e. $51.24

Answer: Move decimal point 1 place to the left and then add 1/2 of the amount.

Investigations
Geometry: Who lives where?

10. If Chelsea lives closer to the airport than Terri does and Terri's estate is between the city and the airport, then which of the following statements must be true? Hint: Assume the points are on a line and draw the possible locations or cases.

a. Terri lives nearer to where Chelsea lives then to the airport.
b. Chelsea lives between the airport and Terri.
c. Chelsea lives nearer to Terri than to the airport.
d. The airport and the city are the farthest apart.

Answer: d

Applications: Vectors

11. We are used to adding quantities the same as we add numbers, but if the quantities involve direction (in science these are called vectors), then the answers can be different. Assume you are traveling north 60 miles, and then east 80 miles. How far are you from the starting point? Give two possible answers, and explain!

Answer: One answer is 140 miles. The answer assuming a straight path is 100 miles. Distance depends on the route or path.

Bridge 2: New Money and Old Mathematics

"Neglect of mathematics works injury to all knowledge."

Roger Bacon

Ramp 2.1
Inductive Reasoning

(Do your selected topics for one evening per week.)

This Bridge will review inductive reasoning, which is one of the methods people use to arrive at conclusions. The weakness, in this type of reasoning will also be pointed out. The Banker's Rule of Seventy-Two will be explained, plus the "games" of the monetary world such as compound interest, mortgages and annuities will be investigated. You will use the mathematics you study in this Bridge the rest of your life. Now is also the time to discuss some of the financial programs you may be involved in and all in the family should know about. (A good time to do this is when the premiums or expenses of accounts are due, or else just do one financial program per week.)

After completing the activities in this section, you will understand inductive reasoning and:

1. Be able to explain the weakness in inductive reasoning.
2. Be able to calculate the approximate time it takes money to double using the rule of Seventy-Two.

In your geometry course you may have been introduced to several types of reasoning and ways for arriving at conclusions: inductive, deductive, direct, and indirect. These are the types you possibly remember. An example of inductive reasoning from geometry is in the following activity.

Activity

Draw five different shaped triangles, as described below. Then measure the angles of each triangle with a protractor and calculate the sums of the angles for each triangle. Copy the following on your paper and complete the information.

Draw Triangles(type)and size of Angles:

- a. Right?
- b. Isosceles?
- c. Obtuse?
- d. Acute?
- e. Equilateral?

What is your conclusion with regard to the sum of the angles of a triangle? Are you certain the sum is always what you have concluded? Do you recall the proof for the sum of the angles of a triangle, based on Euclid's parallel postulate 180 degrees? (SUM OF THE ANGLES OF A TRIANGLE IS 180 DEGREES.) If you don't recall, draw a rectangle and put in a diagonal and notice each triangle has a sum of 180 degrees.

Hopefully you did conclude that the sum of the angles of a triangle is 180 degrees, by measurement and then show that the sum is always 180°.

Definition 6: Inductive reasoning is arriving at a conclusion you think is true for all cases by observing a few cases.

Discovery Activity 2.1:

1. A group of students deposit their savings from summer employment with a local investor.

Student #1 saved $100 and deposited it at 8% interest.

Student #2 saved $500 and deposited it at 6% interest.

Student #3 saved $1000 and deposited it at 9% interest.

2. The investor suggests that each student using a calculator multiply their amount by 1+r where r is the decimal form for the interest.

EXAMPLE: The rate is 8% or multiply by 1.08. **Use $100 as the investment and repeat the operation until the amount is doubled to $200.**

Hint: 100(1.08)=108. 100(1.08) = 116.64. 116.64 (1.08)= 25.97.08) = etc.

How many times will you perform the multiplication in order to have $200? (Actually, you will have 216. So, the answer is 9.)

How many times did you perform the multiplication? _____

How many times did the other 2 students repeat the operation using rates 9% and 10%?

Answers: approximately 8 times and 7 times.

The investor then suggested they divide 72 by their interest rate, and compare their calculated answers. Then write what they observed as a conclusion.

The conclusion is the Bankers' Rule of Seventy-two is to provide a good estimate as to length of time it takes for money to double at a given rate of interest. Years = 72/rate (per year)

Activity 2.2
(Be Selective)

1. Record a possible explanation as to the reason the following statements were made. Ask your friends for assistance, and possibly they have other examples.

   a. It is bad luck to walk under a ladder.
   b. Seven years of bad luck if you break a mirror.
   c. Wind from the East is not good for man nor beast.
   d. Teenagers are not good auto insurance risks.
   e. Calm water runs deep.

2. What are some similar statements or Superstitions common to your community?

3. What are some possibilities for the next 2 numbers to fill the blanks in the following sequences?

   a. 1, 2, 4, 8, 16, ?, ?
   b. 1, 2, 4, 7, 11, ?, ?
   c. -4, 2, -1, ?, ?

Comment: There are several possibilities. Let the others explain their reasoning or pattern to predict their answers.

Ramp 2.2
What is an inductive conclusion?

1. What are some possibilities for the two missing numbers in the following sequences?

 a. 2,3,5,8,13, __, __
 b. 1,3,7,15,31, __, __
 c. 1,1/2,1/4,1/8, __, __
 d. __, __, 6,9,12
 e. __, __,6,1,-4,-9
 f. __, __,3,9,19,33

Answers: (Let the students explain their reasoning in these activities)

 a. 21,34
 b. ?
 c. 1/16,1/32
 d. 0,3
 e. 16,11
 f. 0,1

2. Write the rule or method you used in problem 1 for finding the next number.

Answers: Some possible rules are:

 a. 2n-1
 b. 2n+1
 c. (1/2)n
 d. n+3
 e. n-5
 f. $(2n^2 + 1)$ (n is the number of the preceding term)

3. Using the rule discovered in the Discovery Activity, how long (in years) would it take money to double using the following rates?

a. 3%   b. 12%   c. 18%   d. 36%

Answers: a. 24, b. 6, c. 4,d. 2

4. Check your answers to #3 using (1+r%) and your calculator. Example: a. $(1.03)^{24}$ should equal approximately 1.8.

5. Which percentages required more periods then expected?

Answer: The higher rates

6.What is the weakness involved in drawing a conclusion by inductive reasoning?

Comment: The error is that you are never certain about the general case based on a few examples.

7. How do weather forecasters use inductive reasoning?

8. Do medical doctors use inductive reasoning in diagnosing problems? Explain your answer.

9. What is the rate of interest your family's credit card company charges on overdue balances? (This is usually a per month rate.)

10. Use your calculator to answer a few of the following.

a. 1 divided by 9 = ?
b. 2/9 = ?
c. 3/9 = ?

d. 4/9 = ?
e. 5/9 = ?
f. 6/9 = ?
g. 7/9 = ?
h. 8/9 = ?

Now what do you think the ANSWER is TO 9 DIVIDED BY 9?

11. What is the weakness with regard to a conclusion reached by **inductive reasoning?**

Ramp 2.3
Angle Sum Postulate for polygons

The angle sum for a triangle is 180 degrees, and was used as early as 2000 BCE, based on the idea of 360 degrees in a circle. A half circle, straight angle, would be 180 degrees. More on this later. What is the sum of the interior angles in each figures below?

Hint: Use only Theorem 4 for the sum of the angles of a triangle.

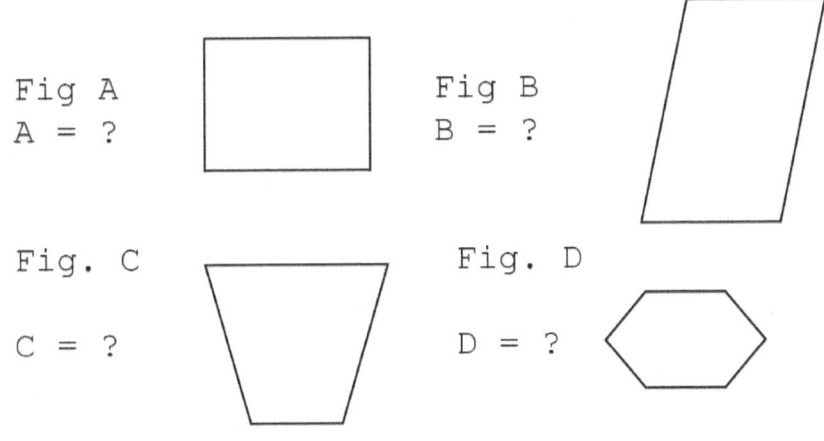

Fig A
A = ?

Fig B
B = ?

Fig. C

C = ?

Fig. D

D = ?

E. What is the sum of the angles in a 6 sided polygon?

F. Draw a seven-sided polygon. What is the sum of the angles?

G. Write a formula for the sum of the interior angles of a N-sided polygon.

Answers: A, B, C, are all Sum is 360 degrees (2 triangles). D. is 720 degrees F. 720° G. s = (n-2)180

Postulate 8: The sum of the interior angles of a polygon is (n-2)180 degrees (Where n is the number of sides.)

What is the formula for the sum of the interior angles in the following figures? (Valid problem and valid conclusion for a concave polygon of n-sides?)

Try several cases and state your conjecture for the formula.

Answer: The formula is valid, but future cases may be invalid since the formula was arrived at by induction!

What is a concave polygon?

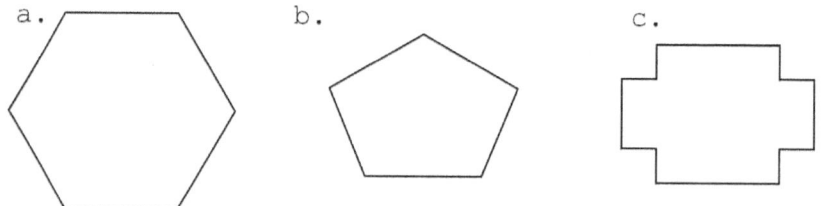

a.   b.   c.

Thinking:

1. A student observed that on a True-False test by a certain teacher, the first question was always

T. The student marked the second question F without even reading the question. What was the assumption the student used? Explain.

Answer: A false question always follows a true question. There could be other responses.

2. Another student observed over the year that a history teacher always gave five T or F questions on the biweekly exam. The student also observed:

   a. That the first and last answers are always opposite.
   b. No three consecutive questions will have the same answers.
   c. There are always more T's than F's.

   The student knew the correct answer to question 2 is F. With a little thought, the student now knows the five correct answers.

   What is the order for the correct answers? Write out your explanation for your answer to part "a."

   Answer: TFTTF

Ramp 2.4
Math Series

"Numbers rules the universe."
The Pythagoreans

Arithmetic Series and Sequences

After completing the activities in this BRIDGE you will better understand:

1. What a sequence is.
2. What a series is.
3. How to determine a specific term in an arithmetic sequence.
4. How to calculate the sum of an arithmetic series.

In Ramp 1 of this Bridge, you reviewed the concept of inductive reasoning, and also the weakness involved in reaching a conclusion resulting from inductive reasoning. Although inductive reasoning is used in everyday situations, we will limit our cases to ones involving arithmetic types. But first, we need to distinguish between a sequence and a series.

Investigation

Below are a few arithmetic sequences and series. These terms are used in the monetary world.

Sequences or Series:

	Next term:
1+2+4+8+16+32	64
2,4,6,8,10,12	??
2+1+1/2+1/4+1/8	??
1,4,7,10,13,16	??
1+3+9+27+1/91	??
20,15,10,5,0,-5,-10	??
20+10+5+2.5	??

Observe the above sequences and series. Then look for the differences and answer the following questions.

1.Which of the following represent a series? a sequence?

      a. 0,6,12,18.
      b. 10,20,30,40.

   c. 18,12,6,0
   d. 3+2+1+0+(-).
   e. 5,.7,.9,.11
   f. 2,4,8,16.
   g. 101+102+103+104.
   h. 1+1.2+1.4+1.6.
   i. a,a+c,a+2c,a+3c.
   j. c+(a+2c)+(a+3c).

2. Give your reason for each answer in question 1.

3. Write your definition for:

   a. Sequence
   b. Series

Answers: 1. a,b,c,e,f are sequences and d,g,h,i,j are series
Answers to "a" are 0,6,12,18 is a sequence.
Answers to "d" is 10,20,30,40 is a Series.

From the above cases, complete the following definitions.

Definition 7: An arithmetic sequence is…
Write your definitions.

Definition 8: An arithmetic series is…

Comment: The group may define the terms to their satisfaction as to correctness, and then enter the definitions in their notebook.

Suggested definitions are:

Definition 9: Arithmetic Sequence: A set of terms such that given the first term, the next one is obtained by adding a constant term.

If the next term is arrived at by multiplication, then it is a series.

Sequences and series can have many terms, and if the number of terms is 100, or 1,000, or a number like the U.S. population, you would object to writing all the terms, and you should. What do scientists and engineers do? They attempt to express the terms of the sequence or series as a formula.

Example: Take the sequence 2,4,6,8,10, ... (The science and business world agree that the . . . means continue in like manner.) So, the above can be written as follows:

Seq.: 2,4,6,8,10, ..., 2n
where n is the number of the term you want the value of.

Questions related to the above **sequence**.

    a. What is the 8th term?
    b. The number 50, is which term in the sequence?
    c. The number 150, is which term in the sequence?
       The number 5002, is which term in the sequence?
    d. The number 71, is which term in the sequence?

Answers: a. 16, b. 100, c. 75, d. not a term.

The problem is to be able to write a formula for the nth term so it is easy to calculate the nth term.

Activity
Work a few – Be selective!

1. Write an expression for the nth term in the following:

---

a. 3, 7, 11
b. 10,0,-10,-20
c. 12, 4,-4
d. 3, 1, -1
e. .5,.2,-1.-4
f. 2, 8, 14
g. 101, 105, 109
h. 1, 1.5, 2, 2.5
i. 0, 3, 6, 9
j. a, a+d, a+2d

Answers:

a. 4(3-1)=11
b. 10-(n-1)10 = -20
c. 12-(n-1)8=-4
d. 3-(n-1)2=-1=
e. .5+(n-1)(.3)
f. 2+(n-1)6 =10
g. 101 +(n-1)4=109
h. 1,1.5, 2. 2.5
i. 0+(n-1)3=9
j. a+(n-1)d

A similar problem exists when working with a series, but it is one degree harder. It is harder since we are looking for an expression or formula to give the sum of n terms. An example is: Sum of 2+4+6+8+10... for n terms.

Ramp 2.5
Number of terms and their sums

Activity

We just add the terms if the number of terms isn't too large. In the case above, the sum for the first five terms is 30. What would you do if the number of terms were 100? Study the following table to help you find the answer for a sequence.

Cases	terms	Term(s)Sum
11	2	2
22	2+4	6
33	2+4+6	12
44	2+4+6+8	20

Do you see a relationship or pattern between the NUMBER OF Terms and the SUM?

Comment: Eventually you will arrive at Sum = (n+1) n, or (n/2)(A+L), where n is the term number and "A" and "L" represent the first and last terms.

Check this method by trying the next few cases.

You can easily see how solving for a formula can be difficult. Another person years ago came up with this method for calculating the sum of an arithmetic series. He or she wrote the series two different ways. (In the following "a" is the first term, "d" is the common difference, "n" is the number of the term, and "L" is the last term.

Think of this as an investment (equation #1):

    Term: T1 T2 T3 T4 T5
       S = A + A+i + A+2i + A+3i + A+4i
       + ... + Last term is A(n–1)I
    So, S = A+(n–1)i

Do you understand the above series? If not go back and read it again.

Another person wrote the series starting with the last term (equation #2):

S = L + L-d + L-2d + L-3d + L-4d + ... +
So, S = L-(n-1)d.

Do you understand the above series?

Now compare equation #1 to equation #2 and arrived at the following sum.

2S = A+L + A+L + A+L + A+L + ...+ A+L
   or 2S = n(A+L)
        S = (n/2)(A+L)

Did you arrive at the above answer when you added the two equations? The person then simplified the expression and labeled it a theorem. "L" is the last term. Do you understand the above formula?

Write a sentence explaining the formula and copy both in your notebook with an example.

Example: A friend loans you $500 at no interest and you plan to pay it back by $100 per month. He said the money is from the checking account which doesn't pay interest anyway. The friend said he will pay $105 June 30, $110 on July1, $115 on August 1 and $120 On Sept.1. Can you imagine solving the problem for buying a house over twenty years, with we're or to the 12 monthly payments.

One way is the using the formula (n/2)(A + L) The total by the formula: S= (4(105/2 + 120/2) which is $112,50 into and if true

Now the case in the 1850s. The money was a loan to a friend who wants to dig for gold in Montana and will pay back the loan with $5 loan cost added on per month. $117.5 is the amount paid each month starting if the payments are to be equal.

June 1: He pays $105 July 1: He pays $110, August 1: He pays $115 and on Sept 1. total payment is $130. Can you imagine the work if this were the case for paying for a house over 25 years with monthly payments?

From the Formula N/2(A+L) we calculate that loan very easily:

$$1/2(A+L) = (1/2)(120 + 105) = \$112.50$$

where n is the number of payments and A is the first Payment and L is the last payment. What is the total of all payments?

Activity

Practice (Work a few cases until you understand)

For each of the following series:

a. Write a formula for the nth term,
b. Calculate the 10th term,
c. Write the sum of n terms and calculate the sum of the first 10 terms,
d. Check your answers by actually adding the terms using your calculator.

1. 6+12+18... Term 10
2. 10+20+30+40+...Term 10
3. 18+12+6+0...Term 10
4. 3+ 2+ 1+ 0+(-)+...term 10
5. 1 + 1.2 + 1.4 + 1.6+...Term 10   9. 0 + 3 + 6 + 9 +... 10. a + a+d + a+2d + a+3d +

Research (Math History)

Who was Carl F. Gauss?

Try a computer search and record for a few of your findings his contributions.

How did Gauss solve the problem for the sum of the first 100 counting numbers?

CASE		
1.	1	Sum is 1
2.	1+2	3
3.	1+2+3	6
4.	1+2+3+4	10

Do you see what he did?

Therefore, the sum of the first 100 is…

Activity: Thinking
Use your notes and calculator. Be Selective!

1. Classify the following as an arithmetic series, an arithmetic sequence or neither.

    a. 1,2,3,4,5,6,

    b. 1+2+3+4+5

    c. -1,-2,-3,-4,-5

    d. -1-3-5-7-9

    e. 2,4,8,10,12

    f. 2+4+8+10+12

    Answers:Seq: a, c

                  Series: b,d

                  Neither: e,f

2. In problem 1:

   a. What is the 15th term in each arithmetic sequence?
   b. What is the sum of the first fifty terms in each arithmetic series?

   Answers: a.15, −15
                 b. 1250, 625, −1250

3. What is the sum of the first 50 odd counting numbers? Complete the following table and look for the pattern.

Number of terms	Terms	Sum
1	1	1
2	1,3	4
3	1,3,5	?

Continue the table until you see the pattern and the sum for the term 50.

Answer: Sum of the first 50 odd numbers is 2500

4. If a person saves one penny the first day, two pennies the second day, three pennies the third day and in like manner for 30 days, how much will the student have saved after thirty days?

   Guess at the amount before you calculate the answer. **(Your friends may enjoy this problem!)**

   Answer: $4.65

5. A way to impress your friends is to tell them you have the ability to add groups of numbers such as the sum of the first 100 counting numbers. The

sum is (100/2)(1+100)or 50(101) which is 5050. Start with some smaller set of numbers like the first ten counting numbers [5(1+10)] = 55 and gradually let them select larger sets. Some suggestions are (try to do these in your head):

a. Sum of the second ten counting numbers,
b. Sum of the counting numbers between 50 and 71,
c. Sum of the counting numbers from 5 to 15, including 5 and 15,
d. Sum of the even numbers between 1 and 101.

   Answers: a. 155 b. 1210 c. 120
            d. 2550

Ramp 2.6
Geometric Patterns

1. Given: an **isosceles** right triangle and each additional isosceles right triangle is constructed on the hypotenuse of the previous right triangle. Calculate the measure of each hypotenuse and predict the pattern for the next few constructions.

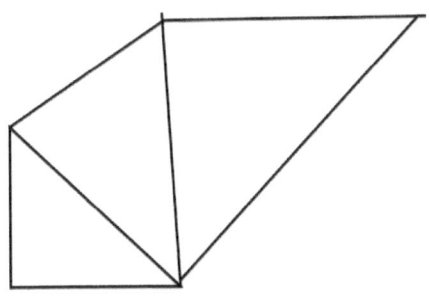

Reading Problem:
(enjoy)

2. How many were going to St. Ives?

Read this 18th century English nursery rhyme (which is a version of an Egyptian puzzle) very carefully, and then answer the question.

> **As I was going to St. Ives**
> **I met a man with seven wives.**
> **Each wife had seven sacks,**
> *Each sack had seven cats,*
> *Each cat had seven kits,*
> *Kits, cats, sacks, wives,*

How many were going to St.Ives?

Problem is from the Rhind papyrus
THE NATURE OF MATHEMATICS
Karl J. Smith

Answer: **Read it very carefully!** Hint: The answer is less than 10.

3. The volume of a pyramid and a cone - an outline for proof!

Draw a cube with a side of 1 or 2 inches to help you see the problem better. Label the sides so the volume of the cube is e^3 cubic units.

Now draw the diagonals and look for the 6 pyramids and you can write Volume of the six pyramids equal $6e^3$. **Therefore the volume of a pyramid is $(1/3)e^3$.**

Explanation!

CUBE=6 (PYRAMIDS)

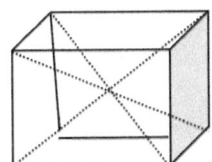

Explanation!
CUBE=6 (PYRAMIDS)

Volume of cube = e^3 with a volume of e^3 and the volume of pyramid is $e^2(e/2)$, or Base area time altitude, but there are 6 pyramids, therefore the volume of a cube 6 times the volume of 1 pyramid.

$V = 3e^3$, or in general: V = B(area) time Altitude(V = B)(h) and 6 pyramids = $6(e^2)(\frac{1}{2})e = 3e^3$ pyramids Bh/3 or = Vol of 1 pyramid where B is the area of the base.

Theorem 7: The volume of a pyramid or a cone is Volume = 1/3 Base area times altitude cubic units.

Investigations: Area relationships

Justify your answers to the following:
Hint: See Appendix 4 for needed formulas

 a. Can the area of a circle ever be the same number as the circumference of the circle? Hint: $2\pi r$ =? πr^2 = ?
 b. Can the area of a square ever be the same number as the perimeter of the square?
 c. A circle is inscribed (just fits in the square, or is tangent to the sides) in a square. Can the area outside the circle, but inside the square, ever equal the area of the square?
 d. A square just fits inside a circle. Can the area of a regular polygon inside the circle ever equal the area of the circle?

Answers: a. Yes, when r is 2.
b. Yes, when s = 4
c. No
d. No, but it can get very close!

Thinking Problem:
Using 5 and 3 to equal 4!

A person has two containers, measuring 3 and 5 quarts. How can they get exactly four quarts of water? (Try to solve this situation before looking at the answer below.)

	Containers	
Answer:	Five	Three
	0	0 both empty
	5	0 Fill the 5.
	2	3 Fill the 3 from the 5.
	2	0 (Explain the rest of the operations)
	0	2
	5	2
	4	3
	4	0

Using some of the investigations in this Bridge, you might choose the topics of annuities or bank loans for your next after-dinner family discussion.

Write your Summary for Bridge 2!

Bridge 3: Money and Mathematics Today

"Foundation of Sciences, and the plentiful Fountain of Advantage to human affairs."

Isaac Barrow

Ramp 3.1
Geometric Series: Money Earns Money

History of money as a topic is very interesting and a topic in the advancements of any country. **There were banks in countries as early as 1000 BCE.**

The activities in this section will help understand and apply:

1. A geometric series,
2. Determine the sum of a geometric series,
3. Calculate the sum of an infinite geometric series,
4. And understand the advancements related to money.

In the last section, you were informed as to the difference between an arithmetic series and sequence. The same definitions are valid for geometric sequences and series, with slight variations. The following are examples of geometric sequences.

This topic was selected since most people are interested in monetary problems and their applications.

    a. 2, 4, 8, 16...
    b. 1, 3, 9, 27...

c. 4, 2, 1, .5...
d. a, ar, arr, arrr,.. or a, ar, ar^2, ar^3,...ar^{n-1}.

From the above examples, complete the following.

A geometric **sequence** is? (YOU COMPLETE).

A geometric **series** is? (YOU COMPLETE).

Compare your definitions with these.

Definition 10: A geometric sequence is a set of terms such that, each successive term is derived by multiplying the previous term by a non-zero positive constant.

(Some people prefer the definition that states the ratio of any two consecutive terms are equal.)

Definition 11: A geometric series is the sum of a geometric sequence.

Record the definitions in your notebook for future reference.

You may recall the Pascal Triangle case from algebra and/or geometry, which illustrates a geometric series for the sum of each row.

Pascal's Triangle Sum

```
        1              1
       1 1             2
      1 2 1            4
     1 3 3 1           8
    ? ? ? ? ?          ?
```

Can you complete the next few lines or rows?

Research: The life of Pascal, his mathematics and his triangle. (See E. T. Bell's MEN OF MATHEMATICS, Chapter 5.)

The problem is to derive a formula for the sum of a geometric series. We will capitalize on the work of former students of mathematics who had derived the formula after some creative thinking. Their method is summarized below. Be sure to understand each step or ask questions.

Given the following geometric series

1. $S = A + Ar + Ar^2 + Ar^3 + ...+ Ar^{n-1}$

Comment: Note the meaning of "A", "n" and "r" then carefully "walk" through the following derivation.

Operation: Multiply the given equation #1 by r. Resulting in the following:

2. $rS = Ar + Arr + Arrr + Arrrr +...+Ar^n$
   or
   $rS = Ar + Ar^2 + Ar^3 + ... + Ar^n$

Operation: Subtract equation 2 from 1 resulting in:
   $S - rS = A - Ar^n$
   and
   $S= A(1 - r^n)/(1-r)$ or $(A - Ar^n)/(1 - r)$

Test the formula in the case a above. $1+2+4+8+16 = 15$
$S = A\{(1-2)^x/-1\} = ?$ (where x, the exponent is this case is 4).

Theorem 8: The sum of a geometric series is:

$$S = A(1 - r^n)/(1-r)$$
$$or$$

$$S = (A - Ar^n)/(1 - r)$$

where A is the first term, r is the factor, and n the number of terms.

(Recall: What is a theorem?)

A special case is when the number of terms is very large and r is between 0 and 1, then re-write the equation in #5 in this form.

$$S = A/(1-r) - Ar^n/(1-r)$$
or another value:
$$1/1 + 1/2! + 1/3! + 1/4! + . .$$

Now if r is between 0 and 1 and the number of terms is very large, then the value of S in #6 approaches A/(1-r). Why? Think about it! What value does the fraction $Ar^n/(1-r)$ approach when r is between 0 and 1 and n is very large?

This leads us to the formula for the sum of an infinite geometric series when r is between 0 and 1 or when the value of the terms (as n increases) approaches zero. The sum approaches a limit or the value of A/(1-r) to the Nth power.

Theorem 9: The sum of an infinite geometric series is:

$$S = A/(1-r)$$
when r is between 0 and 1(0 < r <1)
and n is very large.

Record theorems and include explanations, in your notebook.

Comment: Review the derivation of the theorems over several days to ensure understanding.

Ramp 3.2
Geometric Sequences
Be selective!

1. Fill in the blanks on the following sequences.

    a. 1, 4, 16, __, _
    b. 1/3, 1, 3, __, _
    c. -2, -1, -1/2, -1/4, __, _
    d. -2,1,-1/2,1/4,-1/8, __, _
    e. b, be, bee, beee, __, _

2. What is "r" (the common multiplier or ratio) in each number 1?

3. Write the first 6 terms of each sequence in number 1 as a series.

4. Use the formula for the sum of a geometric series and calculate the sum of the indicated terms in number 1c & d with S = A/{(1-R) to the nth} and R is between O and 1.

5. What is the sum in problems a and b in #1, if the number of terms is infinite?

Answers:

1. a. 64, 256   d. 1/16, -1/32
   b. 9, 27   e. b4, b5
   c. -1/8, -1/16

2. a. 4 b. 3 c. 1/2 d. -1/2

3. Replace the commas with + signs.

4 Formula: $S = a(1 - r^n)/(1-r)$
   a. 1365   c.-3.9375

b. 121.33   d. -1.3125

Activity Thinking
(Be Selective)

Comment: Write the formula for each problem, substitute the values and then solve using the calculator. Suggest you determine the sign for each answer before solving.

1. Continue the following geometric sequences for two additional terms.

   a. 1, 2, 4, 8, __, __.
   b. 8, 4, 2, 1, __, __.
   c. 1/4, 1/2, 1, __, __.
   d. $\sqrt{2}$, 2, 2$\sqrt{2}$, 4, __, __.
   e. 1, 1/3, 1/9, __, __.

2. Write each sequence in #1 as a series.

3. a. Calculate the sum of the first six terms for a few selected series in #2.
   b. Calculate the sum for an infinite number of terms in problems 2b and 2e.

4. Write the sum for the first term in 2c:

   a. Not using your calculator.
   b. Using your calculator.

5. Justify the following:
   $S = 1+1/\sqrt{2} + 1/2 + 1/2\sqrt{2}+...+(1/\sqrt{2})^{n-1}=$
   $2 + \sqrt{2}$

6. A student said the following three numbers form a geometric sequence (32, __, 2). Unfortunately,

the middle term was smeared. What is the missing term?

7. Another student challenged the class with the following problems:

   a. If S = 62, r = 1/2, and n = 5, what is the first term?
   b. If a = 1/16, r = 2, and the last term
   c. is 32, how many terms are there?
   d. If a = 2, n = 3, and the sum is 26,
   e. then what is the value for r and also the value of the last term?

8. **How to be rich in 31 days!!** (You will enjoy this!)

   On the first day, deposit 1 cent in the bank.
   On the second day, deposit 2 cents in the bank.
   On the third day, deposit 4 cents in the bank.
   On the fourth day, deposit 8 cents in the bank.
   On the fifth day, deposit 16 cents in the bank.
   Continue until day 31. (Use your calculator)

   a. What is the amount for the deposit on the thirty-first day?
   b. What is the total of the deposits in dollars and cents?
   c. Where is the weakness in this method for becoming a millionaire?

Some Answers:

2. Replace commas with plus signs.
3. a. 1a.= 31 1b. = 15.25 1c. 7.75
   d. = $14+7\sqrt{}$ or 23.89 (2 decimal places)
   e. 40 $\square$/81 or 1.55 (2 decimal places)
8. a. 230 cents b. S = 2^{n-1} or
   $21,474,836.47

c. The weakness is in the size of the deposits the last half of the month.

Thinking: Where on the Earth?

Where on the earth can you go 10 miles south, 10 miles east, and 10 miles north and be back where you started? Hint: There are many, many places!

(Ask your friends to help you
solve this amazing problem.)

Comment: The first response is usually the North Pole, which is one correct answer. Can you find another answer? There are actually an infinite number of points (places), where this can be accomplished, in theory. One solution: Take a point 10 miles north of the 10-mile circle of latitude near the South Pole as another source of infinitude of points.

Ramp 3.3
Simple Interest
(How money earns money!)
(Need a Calculator!)
(Good for a family discussion.)

After completing the activities in this section, which is probably a review for you, except the Bankers Rule of 72.

What is the difference between?

   a. Simple interest.
   b. Compound interest.

The United States Congress passed the Truth in Lending Act in 1969. It requires the lending institutions to report two pertinent facts when money is borrowed:

1. The rate of interest charged per year, called the annual percentage rate (APR).

2. The total finance charge or what it will cost to borrow the money.

This section will help you to understand how money earns money in the banking business. This money is called interest. It is important to understand the two kinds of interest, simple and compound. You will need a scientific calculator to work the problems in this section and you will find these problems related to the mathematics of finance but the problems can be much harder to solve. Most people, some time in their life, might seek a loan in order to improve their life style, such as for housing, cars, and physical needs. It is a wise habit to save some money and invest it for the so-called "rainy day". It is said that a person will change jobs five to eight times by the time they are 50 years of age. In most cases each change will be a promotion with higher wages and requiring more education or experience. A consequence is the demand or need for continued learning or education.

In the old days the money saved was put away in a safe place, like under the bed or buried in a container. Sounds odd, but it is true. Today people invest their savings in banks, savings and loans, insurance companies, or with investment companies.

This isn't new to you and I am sure you agree that the institution, which has your savings, should pay you for the use of it. This payment is called interest. There are two kinds of interest, simple and compound. Which do you wish to study first? Naturally, the simple interest. You have always started with the simple things first in the learning process and progress to the more complex. That approach won't be changed.

Simple Interest

Definition 12: SIMPLE INTEREST is the money paid or the cost for the use of the money borrowed when calculated using the formula,

$$I = prt \text{ or } I = pr.$$

The formula, $I = prt$, where **I** is the Interest(\$), **p** is the principal (amount loaned or borrowed), **r** the rate of interest, and **t** the time usually in years or the time each party agreed to.

Example 1: John borrows \$300 to buy a printer for his computer and agrees to pay back the loan amount in 4 months at 5% interest. What is the interest and the total payment at the end of the 4 months?

I = pr = 300(.05)= and the interest is $15. For a total of $315.

Depending on the agreement the answer could have been $305. I = P or I = 300(.05) to the 1/3 power and the payment would be $305. The 1/3 is = to 4/12. (1/3 of a year). It is important that both parties agree to the interpretation of the terms.

Comment: The % sign is a symbol for the mathematical number defined as 1/100. Do you see the fraction bar and the 2 zeros in the symbol? This should help you remember the meaning.

In other words, for the use of the money it cost John $5 or $15. Written notes or contracts defining the details such as the names of the people involved, the due date, the rate of interest, the amount of money borrowed, and many times what the money is to be used for may all be carefully spelled out in advance. The note, or contract, is naturally signed by the parties involved. If only one copy of the note is prepared, who do you think will keep the note, the lender or the borrower?

The formula, I = prt, is an equation. Anytime you know three of the four variables of this equation, you can solve for the fourth one.

Activity
Be selective due to your background

1. What is the **simple interest** on a loan of $150 at 6% for 2 years? Suggested method: Write the formula

and then substitute the values for the variables
and solve. Follow this solution:

   I = p x r(%) x (t) (Where the x
   is the times sign.)
   I = 150 x (6)1/100) x 2
   I = 18
   The interest is $18.

2. Calculate the unknown factor in the following:

I	p($)	r(%)	t(Years)
?	178	4.5%	2
?	210 6.2%	2.5	
?	3578	4.9%	5
?	50	5%	4

Answers: $16.02, $32.55, $876.61, 10

Ramp 3.4
Compound Interest

Simple interest is suitable for one-year or less
loans, but as the loan terms became longer such
as a 25-year loan for a house, another method is
developed. This method is called **compound interest.**

**Definition 13: Compound interest is interest not
only paid on the principal but on the interest
also.**

Formula for the total amount is:

   A = P(1 + r)^Y

Where Y = years or periods, A = total amount due (if
a loan) or final balance (if an investment), P = the

principal value of the loan or the beginning value of the investment, and r = the rate of interest

Example: After John paid off the loan for the computer printer and was quite successful, he was offered a new position that required a car. John did not have a car, so he investigated the option of buying a used car. Fortunately, he found a car with low mileage and a valid history, but it cost $5000 and he had only $1000 for a down payment. The bank approved a loan for $4000 with an interest rate of 5.5% **compounded annually with the total amount owed due in 4 years.**

Substituting the values into the formula:

$$A = L(1 + r/)^y$$
$$A = 4000(1 + .055)^4$$
(Where did the 4 come from?)
$$A = \$4955.30$$

At the end of 4 years the amount John had paid the bank is $4955.30.

Postulate 9: Compound interest formula is

$$A = P(1 + r)^y \text{ to the y power.}$$

Where Y = years or periods, A = total amount due (if a loan) or final balance (if an investment), P = the principal value of the loan or the beginning value of the investment, and r = the rate of interest.

What if John wanted to pay the debt by 4 payments?

Loan is paid by 4 equal payments. Loan total equals P1 + P2 + P3 + P4.

The debt is paid in 4 years by 4 equal payments.
4000 plus Interest = L1 + L2 + L3 + L4.

Now, mathematics comes to the rescue to simplify
the problem. To understand the problem, draw a
rectangle with two diagonals as below. Now, add the
first payment to the last payment 1055 = 1238.31

This means John pays 1000 at signing.

**Possibility 1: $1000 at end of first year plus
interest.**

L1 = 1000 + interest 5.5% or $1055
(1000 at end of 2nd year plus interest).
L2 = 1000 + 113.03 or $1113.25
(1000 at end of 3rd year plus interest)
L3 = 1000 + interest = $1174.24
(1000 at end of year 4 + interest)
L4 = 1000 + interest = $1238.82
The total amount is **$4581.31**

Possibility 2: 4 equal payments of $1,145.33

John wondered if there is an easy way to calculate
the 4 equal payments?

The following figure represents the loan.

   Area ADEB = DOE + AOD + EOB + AOB
      Segment DE is the payment plus interest figure.

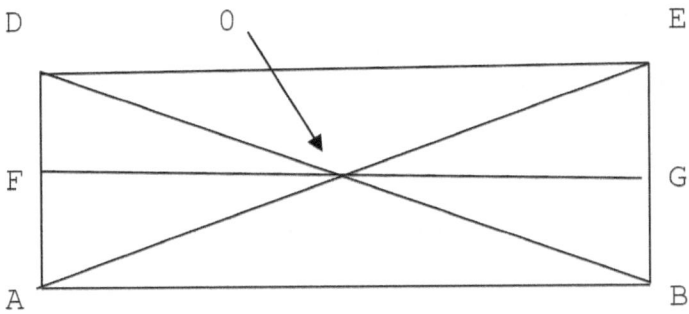

Segment AE represents the Payment plus interest payments and segment DB represents the payments minus the interest payments. Since the Triangles are congruent the small rectangle represents the equal payment amounts line. The triangles are congruent, so we can calculate the four equal payments (over 4 years). Segment DB, the monthly amount, decreases to zero.

Segment FG is the interest plus principal payment line and shows the constant payment amount. The payment line AE increases as the portion of the interest increases. By geometry the triangles are congruent equals, hence the rectangle ABED equals debt and payment amounts, and the pay FG indicates a constant or same payment over the debt period.

The final calculation is T =(N/2)(A + L) (where T is the amount of each payment, N is the number of payments, A is the first payment amount, and L is the last payment amount).

And we can add 1055 and 1238.31 divided by 2 or $1146.66.
T = (N/2)(A + L) is the formula.

Hence, each of the 4 payments are 1146.65 for a total cost of $4586.62. This is an easier way to estimate the equal payments.

Estimated Payment $=(A+L)/2$ times number of payments.

Thinking Activity
Be Selective- work problems
related to your interests.

1. Calculate the compounded amounts (A) in the following:

$L	r%	X	A
a. 2500	6	3	$2977.54
b. 12345	7.5	12	$29403.07
c. 9000	6.5	20	$31712.81
d. 3000	9	8	?

2. A person needs $10,000 in 5 years and invests $8,000 at r% compounded annually. What value will r have to be to result in the desired amount?

Hint: Write the formula, substitute the values for the known values and then solve for the unknown variable.

$$10000 = 8000(1+r/1)^5$$
and take the 5th root of each side
of the equation resulting in
$1.0456 = 1+r$, $r=.0456$ or 4.56%.

3. Which is the better investment? (Guess first. Then do the mathematics.)

a. $5000 at 7% simple interest for five years?
b. $5000 at 6% compounded yearly for five years?

4. Some people think that when simple interest and compound interest have the same r% rate the amounts will be the same.

   a. What do you think?
   b. What will $3000 invested at 7.5% simple interest be 10 in years?
   c. What will $3000 invested at 7.5% compounded annually be in 10 years?
   d. Write your conclusion.

5. What do the Banks and Savings and Loans pay in your community with regard to rate of interest and what are the compounding periods?

6. College tuition plus room and board cost $15,000 in 1999 and if inflation is 3% annually, then what will college tuition plus room and board cost in 15 years? Guess first and then solve.

Answers: 1. a. $2991.70
                b. $30278.88
                c. $32347.81
                d. $32908.02
                e. 3$3019.85
       2. r = 4.56%.
       3. a. $6750   b. $6744.25
       4. b. $5250   c. $6336.19
       6. Hint: Banker's Rule of 72

Complete your notes

The method of estimating debt loan or debt over a period of time is not a simple type of problem. Calculating this value is done using the estimated method shown above.

Ramp 3.5
Family Investments
(Weekly dinner and/or after dinner discussion)

"In the ancient world, as now, trade has been the principal consumer of mathematical preparations."

P. Davis & R. Hersh
THE MATHEMATICAL
EXPERIENCE, page 90

After completing the activities in this section, you will understand better how to:

1. Calculate a Mortgage or Annuity Payment.
2. Define an annuity
3. Calculate the value of an annuity.

The quote above states: the business world is the great user of mathematics. This section will expose you to complicated concepts and applications of mathematics, namely the mathematics for annuities and mortgages. You will likely be in the future a "consumer" of annuities and mortgages as you plan your financial investments.

A GOOD RULE to remember!

Comment: Ask your family banker if their employees use the **Rule of 72** to approximate the time it takes an investment to **double in value?**

Example of the rule: If the rate of interest is 8% compounded, then the investment will double in 9 years. How was the 9 determined? Divide 72 by 8 and the answer is! (This also reminds you that your debt could double in 8 years.)

You may need to talk with some bankers or investment managers' for some of the explanations in this section.

Annuities
(Be selective)

Let's consider the annuity case first, since it is the application of mathematics which most people use first and it will help us understand mortgages. Annuities are realistic ways of saving money and you will be quite amazed at how fast this method will enable you to accumulate a sizable amount of savings. First, what is an annuity?

Definition 14: An ANNUITY is series of a payments, or deposits, made for an agreed amount of time at a specified rate of interest.

Example 1: Let's assume that you are out of school and have your first job, which nets $250 per week (take home). Out of this income, you agree to deposit $3000 and keep it invested for 3 years at 6%. (At the end of 3 years, you will make the decision to stop the deposits, or continue the deposits, or increase the deposits for another specified time.)

In other words, your account looks like this:

	Year	Deposit
End of	Year 1	$3,180.00
End of	Year 2	$3,370.80
End of	Year 3	$3,573.05

The investment interest at the end of the first year is $180. The original investment plus interest at the end of the second year is $3,370.80. What is the total value at the end of the third year?

Answer: $3573.05

Comment: Be sure to "see" where these values came from and verify them.

The above problem isn't very difficult, since there are only a few steps or years involved. What would you do if the term was for 3 years and the payments were made yearly, instead of all at once in the beginning? The answer is, there must be an easier way to work the problem, like a formula where we just substitute values. And this formula would work for any number of years.

If we use the letter **p** for the payment, **n** for the number of periods, **y** for the number of years, **A** for the final amount, and **r** for the annual rate of interest, then we can write a general formula for calculating the **compound interest**:

$$A = P(1 + R/100)^n \text{ (the nth power)}.$$

Let's assume we will invest $1,000 each year, for the following three years, for a total investment of $3,000.

The first $1,000 deposit will earn interest for 3 years or periods.

The second $1,000 deposit will earn interest for 2 periods.

The third $1,000 deposit will earn interest for 1 period.

Comment: **You need to carefully work each time period to determine the amounts, and then add the amounts to get your answers.**

Isn't there an easier way to determine the total amount you are investing plus the interest? Yes, there is an easier way to arrive at an estimated value. (We will use the above problem to illustrate the method.)

    a. The total value of the investment is $3000.
    b. The time of the investment is a total of 3 years.
    c. The average investment, to be made yearly, is $1000.
    d. The interest rate is 6%.

Here is the calculation using the compound interest formula:

    Year 1: $1000 for 3 years is $1191.02.
    Year 2: $1000 for 2 years is $1123.60.
    Year 3: $1000 for 1 year is $1060.00.
    For a total of $3,374.62 at the end of the three years.

But what would you do if your investment was over 30 years instead? That would be a lot of calculating!

Instead, we can use the formula we discussed earlier, which uses the first and last deposits only, but still gives the total at the end of any time period:

$$T = (N/2)(A + L)$$

(Where T is the total amount in your account after the accrual time period, **N** is the number of payments you will make, **A** is the first deposit amount plus accrued interest over the contract time, and **L** is the last deposit amount plus accrued interest over the remaining contract time).

The calculation using the above formula is:
Add: A-First + L-Last Deposits =
$1,060 + $1,191.02
Multiply that answer by N-number of payments (3 in this example),
Then divide by 2.
That gives an answer of $3376.53, which is a good estimate for the value you calculated earlier.

Explanations may repeat and answer questions another day to really understand the method or formula. Many times, especially in math and science, a concept will require several readings before an explanation is clearly understood. A computer spreadsheet may do these automatically.

It is vital to determine if the last deposit is to earn interest or not! This fact changes the number of interest periods for problem and is the key. The problem stated in example 1 is restated below.

Example 1: Let's assume that you are out of school and have your first job, which nets $250 per week (take home). Out of this income, you agree to deposit each year $1000 for 3 years at 6%. At the end of 3 years you will make the decision to stop the deposits or continue the deposits.

The general 1500 case or series in example 1 looks like this:
     (year 1) + (year 2) + (year 3)
of $A = p(1+r)^3 + p(1+r)^2 + p(1+r)^1$.
Or reversing the order since addition is commutative.

$A = 1060 + 1123.60 + 1191.02 = 3374.62$

It is important to note that this is a geometric series. Consequently, we can apply the formula for the sum of a geometric series, $S = a(1 - ar^n)/(1-r)$

p is the payment made each period.
R is the annual rate of interest.
N is number of periods per year.
Y is the number years

This formula will be called the annuity theorem.

Theorem 10: The Annuity Formula for the value of an annuity is:

$$A = p[(1+r/n)^{ny+1}) - (1+r/n)]/(r/n).$$

Working these formulas for monetary related problems is very time consuming so a **similar method** will be used and explained. Reworking the example, which we know the answer to since we worked it out period by period, using the formula for the sum should give the same or approximately the same answer or amount.

$A = 1000[(1.06)^x$ S (S is the Sum
$= \$3334.60)$

The average investment is $1111.53

Observe, this average used the formula
$S = (n/2)(A + L).$

The situation is where you make n equal payments, the same amount at the beginning of each period and the amount is the total at the **end** of the **last** period. This is important, and must be understood in order to use the formula! The following will, hopefully, clarify the procedure.

Say on January 1, you invest $1000 and invest the same amount on the first of the next three months. The total amount, $4000, is left in the annuity for one more month, so that interest will be paid on the last $1000 invested. The rate of interest is 6%. This is why the word "end" was emphasized in the explanation. It would not make sense to invest the $1000, the last payment, and turn around and withdraw the account.

Another way to illustrate or ESTIMATE the answer is the geometric method or the A + L formula discussed earlier.

Deposit: Jan. $1000,
           Feb. $1000,
           March $1000,
           April $1000
withdraw balance on May 1, $4000 plus interest.

In order to easily calculate compound interest we use the formula (S =(n/2)(A plus L) and n in this case is the number of deposits(or payments)in this case. If S had a large number of investments like buying a house where you may have a large number, perhaps 240, payments (20 years for 12 months) Then the average investment or payment is S/N. The final amount including the interest is total $ is (S/N) (1+ R)to y power.

The deposits are 1000 each. Dates: Jan 1, Feb 1, March 1, April May 1.

Substituting the numbers in the formula, will give the same answer. S = N/2(a + 1) = $ = (4/2)(1060 + 1260)= $4640

On May 1 he could withdraw $4637.09. The two answers differ, probably, due to rounding.

Comment: Be sure to understand the conditions to use the formula. **This is not an easy formula to recall and use,** so be sure to record the formula in your notebook, see Appendix 3 with an example and See Theorem 10.

Ramp 3.6
The Mortgage Payment

The formula is different if the situation is for borrowing an initial amount from a bank and you agree to pay it back over a period of time by making equal monthly payments.

The mortgage payment formula basic concept is: Bank loan plus interest (r)= Annuity type payments total (R).

Notice the loan side is the amount of money borrowed at compound interest and the righthand side is the annuity formula. R is the total of equal payments that will be made, which is an annuity type. **The formula really says that the total money borrowed will equal the money paid back plus interest!**

Case: Joe buys a car for $20,000 and will pay it off in five years at 5% interest with a $4000 down payment, so he really owes $16000. The average loan is $8000 for 4 years, or $4000 per year plus interest, or about $18,132.

Comment: You will no doubt have to carefully read and re-read, plus check the algebra, to fully understand and solve the problem using the formula.

See Theorem 10 as in the previous problem. Work the problem 2 ways as in the previous problem.

An easier way to solve the problem is to calculate: the Average Payment method.

Comment: Be sure to understand the conditions to use this formula.

This is not an easy formula to recall and use, so be sure to put the formula in your notebook, with an example! (This is why it was suggested many formulas need not be memorized.) (See Theorem 10.)

Your result should be p (the payment) = $553.79.

Answers: The payment each six months is $553.79 and the total that is paid, including interest, is $3322.75 (6 times 553.79). The interest is (3322.75 − 3000) or $322.75.

Now you need to practice, or work a problem.

Comment: Discuss problems of this type with a local bank loan, especially the number of decimal places they used.

Activity
(Be Selective!)

Hint: First decide which formula fits the problem. Then write the formula and substitute in the given values, finally solve for the unknown.

1.Your parents deposited in an annuity $100 per month for 17 years so that when you are 18 you will have money for college. What is the amount for your college education, if the money was invested at 6% compounded monthly? Do you think this will be sufficient funds for 4 years of college? Guess first and then calculate. (The money is available one month after the last deposit.) **(Always read the contract!)**

Answer: $8874.93 Hint:(T + 1 = 205)

2. A couple plans to retire when the husband is 60. If he is 30 when they began to deposit $2000 per year into an annuity which pays 10% compounded annually and the last payment will be made at age 59, then what will be the value of the annuity when he celebrates his 60th birthday?

Answer: $326,988.05 (T+1=30)

After working very hard on these difficult mortgage problems, perhaps it's time for a break! Let's examine a few "lighter" thinking and logic problems.

Investigations
Geometry: Alphabet Symmetry

Which capital letters of the alphabet, when printed, have:

   a. line symmetry?
   b. point symmetry?
   c. no symmetry?
   d. Are there any letters that are the same when reflected? (This can be done two ways, using a mirror or reflecting over a line as you probably did in geometry.)
   e. Are there any letters that are the same when rotated less than 360 degrees? (All letters are the same when rotated 360 degrees.)

f. Find a picture or an object from nature that illustrates one of the symmetries.

Answers:
   a. Answers may vary depending on how the letters are printed.
   b. One case is O.
   c. F, G, and J are three cases.
   d. I, O, A are a few cases.
   e. I, O, H are three cases.

Application:
Circular Reasoning

What is wrong with this type of reasoning? A student requests the dean's permission to leave early for the day.

The dean asks, "Why?"
The student replies, "To get to work."
The dean then asks, "Why does a student need to work?"
The student replies, "To buy a car."
The dean asks: "Why do you need to buy a car?"
The student responds: "To get to work."

Thinking:
Ten-Coin problem
A Fun Problem
(Try it with your friends)

Arrange ten coins in a row and attempt to rearrange them into five piles of two coins each by the following rules:

1. A jump is defined as jumping two coins and landing on the third coin.

2. Once a pile has two coins, it is out of play, but it can be jumped since it has two coins and counted as a fair move. Your coins now are:

```
       1 2 3  4  5 6 7 8 9 10
Coins: 0 2 3 1/4 0 0 0 0 0 0
```

Example of a move:
Jump coin 1 to coin 4. Now the pile (1 on 4) is a pile of two coins, it is out of play, but the pile can be jumped because it has 2 coins.

The result is:
   0 2 3 (1 on 4) 5 6 7 8 9 10

Now jump coin 1 to coin 4 and then 3 to 5.

Result 2:
   1 on 4 and 3/5 6 7 8 9 10

Notice that coins 2, 1/4 and 3/5 are now "DEAD."

Keep in mind the objective is to have five piles of two coins each. (Notice in the above the objective cannot be achieved since there is no way to move coin 2 to form a pile of two coins. See rule 1.)

Try this problem with your friends!
Hint: Keep track of your moves so you can explain and repeat your solution.

Comment: Challenge your friends to solve the problem. It can be done!

One solution is:
      6->9, 4->1, 8->3, 2->5, 7->10

Record your notes and comments.

Take a break, you have worked hard.

Ramp 3.7
Mathematical Reasoning

"Understanding evolves from work, appreciation is from applications."

Unknown

Review
A must do!

Bridge 3 is the hardest you have had to date. That is why this section is a review. To learn anything new to the degree we understand it, and in turn are proud of our accomplishments, we must practice, practice, and more practice. And then invest something new! In addition, you need to check your notes to be sure they are complete and understandable for future reference. Much-used type of reasoning called Inductive Reasoning.

Inductive reasoning is arriving at a conclusion you think is correct for future cases by observing a few cases. The conclusion is an assumption. Which is the weakness resulting from this type of reasoning, since we don't know if it is true for future cases.

Comment: Arriving at a conclusion with regard to the future from a few past cases is the **weakness** in inductive reasoning. The conclusion or assumption may be false for future cases.

Geometric Example of a false conclusion.: Draw a large circle and follow these directions.

a. Label two points on the circle, A and B.

b. Draw chord AB. How many regions does chord AB divide the interior of the circle into? (2)

c. Label an additional point C on the circle.

d. Draw the 2 additional chords AC and BC. How many regions is the interior of the circle divided into now? (4 regions)

e. Label an additional point D on the circle.

f. Draw chords AD, BD, and CD. How many regions is the interior of the circle divided into?

g. Now complete the table below for the cases you completed above.

Then predict the number of regions when points E and F are added. Check your prediction by counting the regions.

Points	2	3	4	5	6
Regions	2	?	?	?	?

The above problem is credited to Leo Moser in 1950. A beautiful example using inductive reasoning and illustrating its weakness.

Ramp 3.8
Investments and the Rule of 72

(Consider investments as a weekly dinner and/or after dinner topic.)

The second objective was to understand the Rule of 72. This is the **banker's rule of 72** for providing a customer an easy way to predict the years it will take an investment to double.

An example is: Assume you invest $100 at 8% interest. The banker would tell the customer that in approximately 9 years the investment will be

$200. What the banker did not tell was how to arrive at the 9 years is divide 72 by 8.

Use your calculator to answer the following questions.

    a. What is the investment value at the end of the first year?
    b. What is the investment value at the end of the second year? **Continue this process for the 9 years and observe if the banker was correct. The answer is to divide 72 by the rate of interest and the answer is the years to double.**Answers:

a. First year is $108.00
b. Second year is $116.64.
c. Third year is $125.97 and continue the values for 9 years.

In Bridge 2, the following two definitions were listed plus a theorem.

Arithmetic Sequence is a set of terms which given the first term, the next one is obtained by adding a constant term, called the common difference.

Arithmetic Series is the sum of an arithmetic sequence.

$$1 + 3 + 5 + 7 + 9.$$
The sum in the above is:
Terms: 1 3 5 7 n
Sum is: 1 4 9 16 n^2

What is the **sum the first 10** terms? Theorem 5. The formula for the sum is n^2, or in the case above is 100.

The following formula is used to determine the number of equal payments for a payment period like rent or buying a house. The total of the payments including interest divided by N and you have the average monthly payments), P = (A + L)}/n. **Where** A is the first payment, and L is the value last payment, which has been collecting interest all the time This formula method is easier to use then the pattern when the cases are very large, like paying for a house monthly for 20 years More on this later. The following pattern case is easier to use if you see the pattern.

Case	term	sum
1	1	1
2	1 + 3	4
3	1 + 3+ 5	9
4	1 + 3 + 5 + 7	16
50		2500

Do you see the method which is an example of inductive reasoning.

An example using inductive reasoning

When Gauss, a famous mathematician, was in grade school, the teacher gave the class the problem to add the first **100 counting numbers.** (The teacher had some work to do, like making out grades and did not want to be bothered, so the story goes.) Before the teacher had accomplished any work, Gauss had the answer. How did he do it?

a. Let's see if anyone in your group can solve the problem. The problem is to state the sum of the first 100 counting numbers. Who will be the Gauss in your group? Create a table and look for a pattern.

Case #	numbers	sum
1	1	1
2	1+2	3
3	1+2+3	6
4	1+2 3 4	10

Continue until you see the pattern Then state the sum of the first 100 counting numbers!

Let the person explain how they did it.

b. What is the sum using the formula, where n is 100 and A is 1 and L is 100?

Research:
Who was Carl F. Gauss?

Write a biographical sketch of his life.
What were some of his contributions?
How did Gauss solve the problem for the sum of the first 100 counting numbers?
(See E.T.Bell's, MEN OF MATHEMATICS, chapter 14.)

Money Earns Money Review

Definition: A geometric sequence is a set of terms such that multiplying any term n by a non-zero constant derives term n+1.

(Some teachers prefer the definition that states the ratio of any two consecutive terms is equal. See Definition.) **Example: 1, 2, 4, 8, ?.** In contrast, here is the sum of this geometric sequence: 1 + 2+ 4 + 8 = 15

Lemma (What is a Lemma?)
If the number of terms is very large, the result is:

$$1 + 3 + 9 + 27 + ...$$

and $0 < r < 1$, then the formula for the easy sum is:
$S = n(A + L)/2$.
More on this later.

An example of an application using the formula is:

A chain letter informs you to send a letter to each of three selected friends, and they in turn will send one to each of their three other friends, and so the chain continues. (Like a family tree.)

(Many times the letter claims you will have bad luck if you break the chain.)

  a. Draw diagram illustrating the chain letter concept. Put yourself at the first sender.
  b. What is the number of letters mailed at the third sending?
  c. What is the sum of all the letters at the fifth sending?

The mathematics again is difficult, but much more useful in the business and insurance world. The objectives are:

  a. To understand how money earns money in the banking and business world, you need to understand the two kinds of interest.
  b. To understand the types of interest are, simple and compound.
  c. To be able to solve problems related to the mathematics of finance, you need to understand simple and compound interest.

Review of Def.14: SIMPLE INTEREST is the money paid only on the principal.

The formula for simple interest is:

I = $ times R% times T (months or years)
    and T is agreed upon by contract.

(Before banks used this method was used and the contract was many times signed by a third person also. Who else do you think signed the contract?)

Example: Give a friend deposit $100 for 2 years at 15% simple interest each year. The 2 year simple interest total is $30 or $115 and our total is now is $130. Depending on your interpretation of the wording)

The interest rate of 15% seems high but some many credit card companies even charge higher rates and some per month. Ask your friends what their credit card interest rates are, and don't forget your own card.

Definition 11: Compound interest is the case where interest is also paid on the interest. Calculators and computers have made the calculating much easier!

Theorem 12: Formula for the total amount (A) using **compound interest is:**

$A = p(1 + r)^{nY}$ where:
p is the principal.
r is the yearly rate of interest (interest as a decimal rate per year.)
Y is the number of years
n is the compounding periods per year.

Example: Your parents deposit $1000 in the bank for a summer trip 2 years from now. The investment company is paying 15% (compounded annually for the 2 years.)

A very good deal!
$A = p(1 + r)^y$ or $1000(1+.15)^2$
So, A = $1322.50

This doesn't look like compound interest is not much better than the

Try the two problems using $10,000 and 10 years!

Ramp 3.9
Annuities
(Another weekly topic for dinner and/ or after dinner topic!)

This last Ramp is the hardest and yet it may be the most important and useful. You will not be expected to memorize the formulas, but it will test your ability to use the calculator. Banks, insurance companies, and investment brokers have all of these formulas programmed in their computers, but the employees are expected to understand the concepts.

Bankers will be impressed with your understanding of the mathematics of finance after you complete this Bridge and the practice material.

Definition 16: An ANNUITY is a series of equal payments or deposits, made for an agreed amount of time at a specified annual rate of interest.

Theorem 13: Annuity Formula
   $A = p[(1+r/n)^{ny+1} + (1+r/n)]r/n)$
   or
   $A = np[(1+r/n)^{ny+1} + (1+r/n)2/r)]$

where:

A is the value at the end of the contract term,
p is the payment made each period.
r is the annual rate of interest.
n is number of periods per year times the number
of years. (Most of the time n is 1.)

**An easier method, one that you will remember, will
be explained and used.**

An example of an annuity: your kind grandfather was
so elated when you were born, he set up an annuity
in your name for college. He agreed to deposit
$1000 per year for 20 years. The contract reads:
**compounded yearly at an annual rate of 6% for 20
years.** What will the value of this annuity be one
month after the last payment?

The picture is:
    The first deposit is $1000.
    The last deposit is $1000.
    End of year 2, the total is $2247.20.
    End of year 3, the total is $3573.08.
    Year 20 is 20 years of payments and compounded
    interest for 20 years.

In this case:
    p is $1000 (the payment made each period).
    r is 6% (the annual rate of rest).
    n is number of periods per year
        (1 in most cases).
    y is 20 years.
    S is the value at the end of the contract term.
        This is what you need to solve for.

Substitute the given values in the formula:
    A is the first payment and L is the last payment
S = (N){(L + A)/2} and solve.

S = A is $1000(1+.06)^{20}$ and L is$(1+.06)$]
And N is 20
You can easily understand why a calculator is needed and banks have computers!

Answer: A(total) = $32,071.35

Another but easier way to arrive at the answer is to average the total after it is compounded at 6% over the 20 years, which is $32,071.35 and the amount of the last payment which is $1,060. Now, add the first and last totals and divide by 2, which is $1,603.50. The average for the twenty years times 20 is the amount total $32,071.20. (The difference is due to the rounding off method.)

The pay out of an annuity is the reverse problem with a slight change. Take the above case. Now you are ready for college and your contract states you can withdraw equal amounts per month over a 7-year period. How much will you receive monthly? The $3207.12 rate of interest will be 6%. You should understand that the bank has the "A" amount of your money, which means they will or may pay you interest and pay you a specified amount each month. You can see this is the reverse to paying off a mortgage where they pay you an amount of money.

Calculate the amount you can expect each of the four years in college?

In our example this becomes $35,499.72 invested and p is the number of payments. Solving for p, the payment, we have a monthly payment of $518.60 for 7 years. You should verify the $35,499.72 and $518.60 per month for 7 years.

―――

Comment: Answers may vary due to the calculator and rounding. The above answer was not rounded off until the final answer.

1. Interesting problem: Complete the following table.

Number	Square of the number
5	25
15	225
25	?
35	?
45	?
55	?
65	4225

a. Do you see a pattern or method for squaring numbers ending in 5? Try your method for 85 squared and check your answer using your calculator.
b. Did you use inductive reasoning?

Answers:
a. There could be several methods. One method is F(F+1), for 25 where F is the first digit on the left, and then place the product to the left of 25. Example: $25^2 = 625$
b. Yes

2. The Federal Government provides money for funds for various programs at the local level. The money passes through the federal agency, then to the state treasury, then to the state agency, then to the county treasury, then to the county agency, and finally to the local program. At each step 10% of the money is used to cover expenses. If the original amount is $500,000, then what does the local program receive?

Answers: 53.1441% of the $500,000 = $295,245.(Answer will vary some according to calculator and/ or rounding.)

3. Assume you have just graduated, are not married, have a job, and plan to deposit $200 per year at a great rate of 10% over a 3-year period. The money will be left in the bank for one month after the last payment. What is the accrued value? This can be worked by formula or actually per month since the time period is so small.

4. Your uncle leaves you $10,000 to be paid to you in equal installments the first of each year for 10 years. The money is invested at 6.5%. What will the annual payments be?

Answer: $1669.26

Record your notes and comments.

The formulas need not be memorized, but concepts and definitions should be understood. This has been the most difficult Bridge to get over and probably the hardest in this book. **So, have a good weekend!**

Bridge 4: Counting Methods, Probabilities, Odds, and Games

(Plan Your Evening Family Discussions)

"The theory of probability entered mathematics through gambling."

P. Davis & R. Hersh
THE MATHEMATICAL EXPERIENCE

This is another very useful Bridge as to the mathematics for everyday situations. The concepts explained and applied are **permutations, combinations, probabilities, odds, and FAIR BET.**

The topic of a **fair bet** is very interesting and practical. This Bridge will also re-enforce the meaning of the other terms.

Suggested topic: Health Insurance for the one discussion per week night.

Ramp 4.1
Some Algebra, plus Descartes' Gift

After completing the activities in this Bridge, you will be able to:

1. Apply and understand what a mathematical permutation or combination for a set are.
2. Calculate the number of permutations or combinations of a given set.

The theory of probabilities is, as mathematician Laplace stated, "only common sense reduced to calculations." (Who was Laplace? This is another opportunity for an interesting bit of history.)

It makes us appreciate with confidence what reasonable minds feel by a sort of instinct. It is not surprising that the science of mathematical probability originated in the games of luck.

The chance of winning and the intuitive feeling for a winning event leads to the search for easier ways of determine the winning edge. This section will be devoted to the first Ramp associated with the search as to ways for counting the possibilities.

To begin with you will need a few definitions to facilitate communication. Like any game you have to learn the vocabulary. The first term is permutation.

What is a Permutation?

The normal procedure for a student who encounters a word they are not sure of its meaning is to use the dictionary. Your dictionary may provide the following definition.

Definition 17: Permutation: A change from one state, order, or position to another.

This definition is not too clear unless examples are given to illustrate the meaning. (Do you realize that the dictionary attempts to do an impossible job, which is to define all words. List a few words that are impossible to define.)

Comment: With your friends, select from one of their books a sentence or paragraph at random and classify each word as definable or non-definable. Interestingly, the results will be about 50-50. In other words, half of what we say or read is undefinable!

In mathematics and the business world (and also in life situations) a valid definition is concise, understandable, and is true when **reversed**. You learned in your geometry class the **condition of reversibility** for a valid definition. An example of an invalid definition is the following student's definition of a dog, since when reversed it is not always true.

Definition: If it is a dog then it is a four-legged animal. When reversed: if it is a four-legged animal then it is a dog. False

A better definition for a Permutation is:

Definition 18: A permutation of a set of objects is a different arrangement of the objects in the set.

Examples:
a. If the set consists of the letters AB, then a permutation of the set is BA.
b. If the plan is to travel from Chicago to Tulsa to Denver stated as (CTD), then one permutation is to travel from Chicago to Denver to Tulsa (CDT).

The concept of permutation doesn't appear to be so difficult, but in mathematics we are concerned with the possible NUMBER of permutations or the number of ways the set can be rearranged. This appears simple but the cases can and will become more difficult as the applications are discussed.

Try to discover the formula for the number of permutations for the following.

a. How many ways can you arrange the letters AB?

b. How many ways can you arrange the letters ABC? Ans: 6 Write them.

c. How many ways can you arrange the letters ABCD? Write them down.

d. Complete the following table and try to see the connection between the number of letter and the number of permutations. Write the formula for the relationship of the number in the set.

	Letters	Permutations
AB	2	2
ABC	3	?
ABCD	4	?

Do you see a pattern?
Can you predict the following case?

N objects
Answers: 2, 6, 24, N!
Do you understand?
What does N! mean?

Postulate 10:

The number of permutations of n different objects taken all at a time is n factorial, also written as n!

(Read the next paragraph for the meaning of n!)

In mathematics, we are always looking for easier ways to write answers or descriptions such as formulas. Many times, the symbols are dictated by the known symbols on the typewriter or computer keyboard or the printing press. In this case, since the numbers are multiplied, the term selected is derived from the word factors. (Factors are numbers, which are to be multiplied.) So, the mathematicians used the symbol "!" after the positive integer (n!) to indicate an abbreviated way of writing the meaning or the product.

Examples: 3! = 1x2x3 = 6
          5! = 1x2x3x4x5=120

You can see this really is space saving, especially as n gets larger. **Does your calculator have the symbol (!) on it?**

Definition 19: N! read as N factorial, means to multiply all the positive integers from 1 to N and including the integer N.

Examples:
    5!   = 1x2x3x4x5 = 120
    10!   = 1x2x3x4x5x … x9x10 = 3628800
    N!   = 1x2x3x4x5x … x(N–1)x N

Ramp 4.2
Permutation Activities
Be Selective

1. Complete the following and use your calculator if needed.

        a. 7! = 5040
        b. 10! = ?
        c. 15! = ?
        d. 10!/5! = ?
        e. 6! x 3! = ?
        f. 10!/8! = ?

In the above, it was stated that permutation problems would become harder as you investigate them. This always seems to be the case as you investigate any

Mathematica concept. Take the set ABC. How many permutations or arrangements are there, if you take only two elements at a time? List them, count them, and look for a formula.

2. Do the same for the set ABCD, taking 3 letters at a time.

   Answer: ABC, ACB, BCA, BAC, CAB, CBA, ABD, ADB, BAD, BDA, DAB, DBA, ADC, ACD, CAD, CDA, DAC, DCA BDC, BCD, CDB, CBD, DBC, DCB

   Total is 24 or 4!

3. Try this problem, which is similar to the one above. Take the set ABCD. How many permutations or arrangements are there if you take only two elements at a time? List them, count them, and look for a formula.

   Answers: AB, BA, AC, CA, AD, DA,
   BC, CB, BD, DB, CD, DC.
   Total is 12 or 4!/(4-2)! or 12.

Postulate 11:
The formula for the number of
permutations of n different objects
taken r at a time is P(n,r) is n!/r!

There are more cases to be considered. What if all the objects are not all different? Take as an example the case of arranging the letters in the word EYE. Notice two of the letters are identical. List the ways, count them and look for a formula.

Answers: EYE, YEE, EEY or 3!/2! = 3

4. Do the same for the word TOOT. Answers: TOOT, OTOT, OTTO, TOTO, TTOO, OOTT or 4!/(2!2!) or 6.

Postulate 12:
The formula for the number of permutations of n objects taken all at time where r objects are the same, s are the same, and t are the same, etc, is n!/(r!s!t!).

Examples:
      a. Do cases 3 and 4 above.
      b. For the word ROTOR?

      Answer: b. 5!/(2!2!) = 30

      How many permutations are there for the name of your state.

Ramp 4.3
More Permutations and Possibilities
(Be selective)

In problems 1-10, evaluate the expressions. (Answers are listed after problem 30. Use your calculator. If there are 5 persons, then each would have 5 problems)

1. 4!
2. 5!
3. 6!
4. 7!
5. 8!
6. 25!/24!
7. Why do the answers to 5!, 6!, 7!, and 8! all end in zero?
8. What is the smallest value of n where n! end in 2 zeros?

---

9. 6!9!/3!7!) This is a miserable way of writing which integer?

10. If the national debt to Social Security is approximately $4.5 trillion then this is between which values of n expressed as n!

11. What is the largest value of n for which your calculator will give an answer for n!?

In problems 12-17 use the formulas.

12. If a combination lock has ten digits on the dial and three numbers are to be dialed to open the lock, then what is the total number of possibilities?

13. A car dealer can provide the following for each new car:
    a. five exterior colors
    b. three types of interior colors
    c. two types of engines
    d. two types of drives (shift or automatic.
    e. three body styles

    Question: How many cars would the dealer have to order to have one of each possibility? (make up a situation)

14. A small states has the requirement that each vehicle license plate must have 4 symbols on it. The four symbols can be selected from the alphabet (caps only) or the ten digits in any order, but no repetitions. What is the total number of different plates? (Zero – 0 – is different than letter O.) Are there any letters you would recommend not be used? Are there any types of words you would prohibit?

15. Five cards are drawn from a deck of 52 cards. How many possibilities are there for the five cards?

16. A pair of dice is tossed. What is the possibility that you will get:
   a. 5 and 1?
   b. 3 and 5?
   c. 6 or 4?

Answers:
   1. 24
   2. 120
   3. 720
   4. 5040
   5. 40320
   6. 39916800
   7. There is a 5 and a 2 or 10 in the set of factors.
   8. 10
   9. 8640
   10. 12! And 13!
   11. Depends on calculator, but is usually 69!
   12. 1000
   13. 5x3x2x2x3 = 180
   14. You may wish to review the tree diagram method for solving this type of problem. 6 is the answer.
   15. a. 58905 b. 0, O, Q c. profanity
   16. a. 1/36 b. 1/36 c. 18

Ramp 4.4
Geometric Diagonals

1. Which of the following are true?

   a. If the diagonals of a quadrilateral are equal then the figure is a rectangle.
   b. If the diagonals of a parallelogram are equal then the figure is a rectangle.

c. If the diagonals of a quadrilateral are equal and bisect each other, then the figure is a rectangle.

Answers to #1: a. F b. T c. T

Applications

2. Why does a 4-legged chair sometime wobble?

Hint: A geometric plane is determined by 3 points.

Ramp 4.5
Combinations

After completing the activities in this section, you will understand:

1. The definition a combination.
2. How to calculate the number of combinations for simple sets.

In Ramp 4, it was explained what a permutation is and how to calculate the number of permutations in a given situation. It is vital for you to understand the difference between a permutation and a combination. Once you understand the difference between a permutation and a combination, then the formula for calculating the number of combinations of a set will be explained and appreciated.

Definition 20: A COMBINATION is the way a set of objects can be shown where order does NOT count. (AB is the same a BA.)

Comment: Compare this definition with the definition for a permutation.

Case 1: Write all the possible combinations for the word CAT.

Notice the definition says order does not count. This means the arrangement CAT is the same as CTA, or as ACT, or as ATC, or as TCA, or as TAC. If they are all the same, then there is only one combination for the letters CAT taken three at a time.

Case 2: Write all the possible combinations for the word GOLF.

Case 3: Write all the possible combinations for the word SCHOOL.

Comment: The answer to each in cases 1-3 is one. Do you understand why? Now try these cases.

Case 4: From five students (named A, B, C, D, & E) how many ways can two person teams be formed? List all the possible teams.

This is a combination problem since the team consisting of A and B is the same as team BA.

Answer: 10

Case 5: A track coach has five girls who can run the 100-meter dash in approximately the same time. If the coach can only select two runners to run the dash event, then how many ways can he select the two runners? List the possibilities. (Do you see this is the same problem as case 4?)

Answer: 10

You are probably hoping there is a formula to make the calculations easier, and you are correct. The formula will be stated as a postulate since it will not be proved.

Postulate 13:

The number of combinations of n different objects taken r at a time is: $C(n,r) = n!/r!(n-r)!$

Use the formula to verify your answers to cases 4 and 5 above!

So using the above equation for Case 4 and 5: Substitute $n = 5$ and $r = 2$ in the formula: $C(n,r) = n!/r!(n-r)! = 5!/2!(3!) = 1 \times 2 \times 3 \times 4 \times 5/1 \times 2(1 \times 2 \times 3) = 10$

This a case, as I mentioned in the advice to the student, where formulas need not be memorized.

Comment: There are other ways of writing $C(n,r)$ as you will observe, if you look at different books under the topic of combinations.

Activity
(Be Selective!)

Use the formulas and your calculator, when needed, for the following. No need to work all of them.

1. $C(3,2)$ 2. $C(4,1)$ 3. $C(4,2)$ 4. $C(4,3)$
5. $C(4,4)$ 6. $C(5,4)$ 7. $C(5,1)$ 8. $C(5,2)$
9. $C(5,3)$ 10. $C(5,5)$ 11. $C(4,5)$

12. How many committees of 3 students can be selected from a group of 10 students?

Answers: 1. 3 2. 4 3. 6 4. 4

---

5. 1 6. 5 7. 5 8. 10
9. 10 10. 1 11. Impossible
12. 120

Comment: Problems 5 and 10 will require an explanation for 0! **0! is defined as 1 in order for the answers to problems 5 and 10 to come out correctly.**

Activity
(Be Selective!)

1. A teacher gives a test consisting of 10 questions and instructs you to select 3 questions for the test. How many ways can you select 3 questions?

   Answer: 720

2. A basketball coach has 10 players. How many different teams of 5 players could be selected?

   Answer: 252

3. Which do you predict is larger C(10,6) or C(10,4)?

Check your answer. Or are the answers the same?

For the next problems, evaluate 4-->7.
4. C(70,5)
5. C(85,85)
6. C(75,71)   Answers: 4. 12103014
                                        5. 1
                                        6. 1215450

7. A combination lock has 40 integers on it (1--
   >40 and is opened by turning right, then left
   and then right and stopping at the correct
   numbers. What is the total number of possible
   combinations?

Answer: 40x40x40 or 64000.

(Discuss the question with your friends: Should the locks really be called "permutation locks?")

8. Are telephone numbers permutations or combinations?

Answer: Permutations

Application: Your Image
Thinking: The Dating Problems

If three girls (named A, B, & C)
and three boys (named X, Y, & Z)
agree to dance with each other at a party,
then list all the possible
combinations for the six to dance.

Record your notes and comments.

Bridge 5: Equations and Lines

REVIEW Of Some BASIC ALGEBRA
And Descartes' Gift

Ramp 5.1
Equations And Formulas

Bridge 5 will start again with definitions and point out the difference between a formula and an equation. The postulates for solving equations are listed. Descartes' gift, as to graphing equations, is reviewed and extended. The equation for a line is analyzed, so you will understand the meaning of the terms slope and intercept. The useful concept of Direct Variation is defined and applied. (RECALL)

"Students of mathematics...the first time something new is studied they seem hopelessly confused! Then, upon returning (to the concept), after a rest, everything has fallen into place."

E. T. Bell

MEN OF MATHEMATICS

When you complete this section, you may recall your high school days. Can you recall the difference between a formula and an equation?

Probably not, so this is where we will start. You no doubt recall the concepts of equation and formula from your previous work. Perhaps your recall isn't complete, but as the quote above implies the second or third time around, the solutions become less difficult. How would you define the following?

a. Formula   b. Equation

Students usually ask the question: "Isn't there an easier way to work the problem?"

The following example illustrates the significance of the question and to perhaps impress upon you that all students go through the same experience.

In fact, the mathematics textbooks used today are really compendiums of the improved methods and answers created by persons who contemplated the question "How?". The method of solving problems was first by trial and error.

Problem: Bob has twice as much cash as Will has. If Bob gives Will one dollar, then they have the same amount. How much did each have origins sly?

Try to solve the above by trial and error. Do you find a way to quickly narrow in on the answer?

Comment: Let the family members, or persons in your group, express their thoughts and answers. Someone will probably solve it quite rapidly.

Another way to solve the problem is to use algebra. Each sentence in the problem can be translated as: $B = 2W$ where B is Bob's amount and W is Will's amount.

> $B - 1 = W + 1$
> By substitution we have
> $2W - 1 = W + 1$ and consequently W
> $= 2$ and $B = 4$. Translating again
> this tells us: Bob has \$4 and
> Will \$2.

In a way, you might say using algebra takes the guesswork out of the problem!

The word "algebra" comes from the Arabic "al-jabr" meaning "restoration", and refers to the fact that what you do to one side of an equation you must do to the other side to "restore" the equality. This brings us back to the words you were asked to define: formula and equation.

Definition 21: An equation is a statement indicating two numbers are equal.

Example: a. 4 = 16/4 or 4 = 4.
           b. 3x – 7 = 2x + 3
           (In example b. above, x = 10)

Definition 23: A formula is an equation representing a general rule, such as a theorem.

Example: The formula for the area of a circle is πr^2 or $A = \pi r^2$.

The postulates for solving equations are:

Given an equation, you can add a number to each side of the equation.

Comment: This permits subtraction since subtraction is adding the opposite, and also the postulate includes algebraic expressions since they represent numbers.

Postulate 15: If given an equation, then you can multiply each side of the equation by the same non-zero number. (This postulate includes division, powers and roots.)

Comment: Dividing is the inverse of multiplication and dividing by zero is "out" since 1/0 is not a number. Can you justify that 1/0 is not a number?

Ironically, the solving of the equation is not the difficult part. Arriving at or writing the equation is the most difficult and what most students dislike. The process of writing the equations is the most important.

Naturally, we will start with the easiest types first to build your confidence.

Activity
Be selective

1. Which of the following are equations or formulas, both or neither?

    a. $X = 5$
    b. $A = bh$
    c. $2x + 1/2 = 60.5$
    d. $D = rt$
    e. $6/(x-2) + x/4 = 3$
    (X can not be 2. Why?)
    f. 4 inches = 1/9 of a yard

Answers: a, b, c, d, and e are equations, b,d are both, and f is neither since 4 is not equal to 1/9, but it is still a true statement.

Comment: In the above x cannot be 2 since the denominator would then be zero, and remember from Bridge 1 we showed division by zero is impossible.

2. Solve the following equations using the postulates. (Some may not have answers. (Select 3)

    a. $3x + 2 = 3$ b. $y + 2 = 9$
    c. $x - 5 = 7$ d. $3z + 2 = 3z$
    e. $2t/3 = -4/5$ f. $0.4x = 50$
    g. $4(2x - 3) = 3(2x + 4)$
    h. $x\sqrt{2} = 3\sqrt{5}$ (use calculator)

Note: $\sqrt{}$ is translated as the positive square root.

Example: The symbol $\sqrt{4}$ is 2, but the square root of 4 is +2 or −2.

   Answers: a. 1/3 b. 7 c. 12
    d. no answer (not an equation)
    e. −6/5 f. 125 g. 12
    h. $(3\sqrt{10})/2$ or 4.74

3. Follow the solution to this problem. The hard part is to write the equation. At the county fair, Ann won $64. She paid her parents the $28 she had borrowed. She deposited in the bank $10 more than she kept for spending money. How much did she keep for spending money? How much did she put in the bank?

Equation: $64 = parents paid + bank $ + spending so the equation is $64 = 28 + (s+10) + s, where s is the money, she keeps for spending money.

(Do you understand this equation?) Now solve for s, the spending amount.

   Answer: s = $13, Ann deposited $23.

4. Check your answers to #2 and #3 by substitution.

    Example: $2x - 6 = 8 \longrightarrow x = 7$

    Check: $2(7) - 6$ should equal 8
               $14 - 6$ should equal 8
                    $8 = 8$ checks

Activity
Be Selective

1. Solve the following equations (select two):

a. $3x/7 = 6/28$ b. $23y = 46$
b. $5x\sqrt{3} = 9$ d. $y - 12 = 3/y$

2. There are 1260 students at the college. The ratio of men to women is 3 to 2. How many students are women?

    Answer: $W = 504$

3. Angle A is seven times angle B, and angle B is 1/4 angle C. Hint: What is the measure of each angle in the triangle? Hint: Write 2 equations!

    Answers: $A = 105°$, $B = 15°$, $C = ?°$

4. Sue has a test average of 77 on three tests, where all tests have 100 as the perfect score.

    a. What will she have to score on the 4th test to have an overall average of 80?
    b. Can she bring her average up to 85 with a fourth test.

    Answer: a. 89 b. impossible

5. Messenger A is traveling at 55 miles per hour. Messenger B notices A has left an important package and leaves 1 hour later at 75 miles per hour to overtake messenger A. If A left at 9 a.m., what time will B overtake A?

6. Draw a **picture or figure** related to each of the following **formulas** and indicate what the letters or variables represent. (See a geometry text, if help is needed.)

a. $A = (1/2)bh$ b. $P = 2(L+W)$
b. $C = 2\pi r$ d. $V = LWH$
c. $A = (h/2)(b_1 + b_1)$

Ramp 5.2
Conjectures and Justifications

1. Draw an acute scalene triangle and bisect each angle. Label the triangle ABC and the point of intersection of the bisectors label 0.

Draw a line XOY parallel to AB where X is on AC and Y is on BC.

a. Measure segment XY.
b. Add the measures of segments AX and BY.
c. What is your conclusion for n?
d. Can you Justify your conclusion?

Answer: XY = AX + BY. Hints: Mark the equal angles. What kind of triangles are formed? What do you know about these triangles?

Application: Sound or noise

---

2. Take a rectangular container (shoe box) with an open top about 4 by 8 inches and stretch 4 or 5 different size (thickness) rubber bands around it. Now pluck each rubber band as you would a guitar and listen to the sounds.

      a. Write as many conclusions as you can.
      b. Repeat the experiment with a deeper and/or shallower container.
      c. Write any additional conclusions.
      d. How would you define noise?
      e. How would you define music?
      f. Noise may be defined as any sound displeasing to the ear.

Odd questions and answers

1. What is ½ of 8?
2. What is ½ of XII?
3. Which is larger 9 or 3?

          Answers: 1. zero or four or 3
                    2.   7 or 6
                    3.   9 or 3
      (Depends on interpretations)

YOUR summary:

Bridge 6 The "Game" of Solving Equations

"...the essence of plane analytic geometry lies in the matching of ordered pairs of real numbers with the points on a plane."

E. Kramer
The NATURE AND GROWTH OFMODERN MATHEMATICS

Weeknight after dinner, suggest a discussion on the family investments.

Ramp 6.1
Solving Linear Systems

RULES FOR THE 'GAME'

As you may guess there is more than one way to solve systems of equations and inequations. Mathematicians are always looking for easier ways to solve problems. The advances in technology have played a major role in this area. Carefully read and understand the definitions, postulates and theorems, resulting from a group of philosophers, two were Socrates and Plato, since these are the keys to success. It will be repeated many times that the game of geometry is key in the effort to create a program to teach young people how to arrive at logical conclusions and become leaders. This was so impressive that it became adopted by European leaders and even today is required by colleges and high schools. **Fawcett, at Ohio State University in the 1930s, wrote the NCTM yearbook, Nature of Proof, which justified that geometry does teach critical thinking if the**

course is taught using current every day decision making applications.

All conclusions are based on undefined terms, definitions, postulates and theorems or conclusions (laws). (Read the first few lines of Jefferson's Declaration of Independence as a group activity.) This is how all geometry texts are written, undefined terms, definitions, postulates and theorems.

What is a theorem?

What would you say the LAWS are based on?

What are the 4 forms of an implication? How are they related?

In the previous Bridge, you practiced a few problems by solving systems of equations and inequations by graphing. The graphing method is slow and the solution point values are hard to read from the graph unless you have a graphing calculator. The graph provides a picture or a common sense interpretation for the problem and a "ballpark" location for the answer.

After reading, asking questions, understanding the method, and working some of the activities in this section, you will understand the solution methods.

1. Solve equations and inequalities.
2. Interpret the solutions.

Like learning to play a new game, you have to read and understand the rules. It is the same in mathematics. Some people even classify mathematics as the most important game considering its role in and contributions to our quality of life. So as a review, we will start with some definitions and postulates. The major objective for this section is for you to understand the "rules."

A formula is an equation stating a valid relationship, or a theorem.

Example: A = bh for the area of a parallelogram.

Postulate 16: If given an equation, then you can add (or subtract) a number to each side of equation.

Postulate 17: If given an equation, then you can multiply each side of the equation by a non-Zero number. This includes powers and roots.

Definition 24: A point is a set of ordered numbers writtenas(x,y).

Definition 25: A line is defined as the set of points that satisfy the equation $y = mx + b$.

Definition 26: Direct Variation. If two variables are so related that for each value where k is the constant, then x and y vary directly. (K is equal to the ratio y/x or $Y = Kx$.)

We also need a few new "rules."

Postulate 18: If given a set of equations, then the equations can be added or subtracted) and the result is an equation.

Postulate 19: If given an inequality, then a number can be added or subtracted to each side of the inequality and the result is an inequality of the same order.

Postulate 20: If given an inequality, then the inequality can be multiplied (divided) by a positive number the result is an inequality of the same order.

Do you understand what "same order" means?

Postulate 21: If given a set of Inequalities of the same order, then the qualities can be added and the result is an inequality of the same order.

Notice as stated in the former postulate an inequality can be multiplied by a positive number. What happens if you multiply each side by the same negative number? Try a few cases. Write a conjecture for the general case and then give an example for your conjecture. Finally, write the conclusion as a theorem.

Theorem 16: If given an inequality, then the inequality can be multiplied by a negative number, and the result is an inequality of the opposite order.

Like any sport, you need a lot of practice to really play the game well!

Ramp 6.2
Postulate and Theorem Activities
Be Selective

In the following, select a few, and explain which of the above postulates are used to transform the statement in A to the statement in B.

A	B
1. $5x + 2 = 10$	$15x + 6 = 30$
2. $-.5x + 6 < -10$	$x - 12 > 20$
3. $25x - .5 < 3$	$x < 14$
4. $1.2 x - 16 = .7x + 6$	$x = 44$

Answers: 1. Postulate?
2. Postulate?
3. A Theorem and Postulates?
4. Postulate?

Solve the following for X.

In statements 1-4 list the postulates you use to convert each so that the left side is 1x. Solve for x in exercises 6 - 11 inclusive. (select one)

1. $5(2x - 7) - 5(x - 2) = 3x - 41$
2. $2(5x-7) - 5(x - 2) > 3x - 14$

Answers: 1. $x = -8$
2. $x > -5$

3. The cost of gasoline varies directly by the number of gallons purchased.
Using one of the Definitions above, calculate the constant of variation if

7.6 gallons cost $11.97.

$$C = K(\$P)$$

What is another name for the value of k?

> *Answer:* Price per gallon and
> k = \$1.175

4. The coach came in Monday morning to algebra class and said we won by 3, but we out scored the other team by twice their score plus 2 points. What were the scores for each team? What do you think the sport is?

> *Answers:* Scores: 4 and 1.
> Sport possibly is soccer.

Geometry: Conjecture and Proof

1. Do you think the following statement is true or false?

    The perimeter of any quadrilateral is greater than the sum of its diagonals.

    a. Test your answer by measurement using a few cases.
    b. Prove your conjecture is valid or invalid. P > D1 + D2

Application: Air phone

1. Take 5 sheets of 8.5 x 11 pieces of paper and tear each in half lengthwise. Roll each half to form a 1-inch diameter cylinder and tape it. Tape together the 10 cylinders to form one long cylinder. (Cardboard tubes work great if they are available.) Now whisper into the end of the cylinder while the other person listens at the

other end of the cylinder. Let the other person whisper and you listen.

Explain: Why you think the words can be heard so plainly. (This system is still used on ships. Naturally, they do not use paper cylinders.)

Thinking:
Which Bank, A or B?

2. Bank A charges $3 service charge per month plus $.10 for each check you write. Bank B charges $4 service charge per month plus $.06 for each check you write.

    a. Explain under what conditions you should use bank A.

    b. Explain under what conditions you should use bank B.

    c. Explain under what conditions it doesn't matter which bank.

Answers

1. Write four inequalities where one side is a diagonal. Then add them and divide by 2 to arrive at the conclusion. Perimeter is greater than the sum of the diagonals!

2. The sound waves are channeled and cannot escape, consequently, can be heard very clearly.

3. Use bank A, if less than 25 checks, and bank B, if more than 25 checks. If you write 25 checks, the cost is the same at either bank.

Notes:

Ramp 6.3
Solving Linear Systems, Addition Method

"Mathematics seems to endow one
with something like a new sense."

Charles Darwin

In your mathematics courses to date, you probably recall equations with one or two variables. This recall understanding skill is needed and applicable to any profession or vocation you wish to enter. Below are a few examples of formulas used in many occupations.

$A = lw$	Area used in landscaping or decorating
$V = lwh$	Volume used in air conditioning and heating
$D = rt$	Distance equals rate x time, relationship for forms of transportation
$A = p(1+r)^n$	Compound interest formula used in the business world
$S = .5gt^2$	Distance an object falls in (t) seconds - aeronautics
$T = 2\pi(L/g)^{.5}$	Pendulum clock formula engineering $W = VA$ Formula for DC> electrical power engineering
$W = VAcosA$	Formula for AC electrical power - engineering
$1/F = 1/d + 1/I$	used in Optometry
$F = V_1 f/(V_1 - V_2)$	Doppler sound principle - weather forecasting

This is probably enough to convey the idea that equations are used in everyday life but we just don't realize it. This Bridge is devoted to the

147

various ways to solve systems of equations. You will be able to select the method that you feel is the most efficient or the easiest for the problem. In order to select a method, you need to understand each type.

The addition method, as you would guess, employs the operation of addition and the use of the postulates, mentioned above, which basically say that whatever operation you perform on one side of an equation you also perform on the other side. What are the postulates involved?

Why do we need these ways to solve systems of equations? The answer is to have easier ways to find precise answers instead of approximate answers, which you have been reading off the graphs. The key is to convert the problem to one you know how to solve. How is the value of x calculated?

Just substitute the value of y into either of the original equations and solve for x.

$$2x + 4(-1/3) = 7 \text{ ---> } x = 25/6$$

    a. Why is the answer not written in decimal form?
    b. What is a decimal answer in point form?

Comment:

    a. The decimal answer in many cases is not exact since it may be rounded off.
    b. Point form: (4.4,+4.2)

Given: $3x + 5y = 12$
                 $2x - 3y = 9$
                 Solve for x and y.

The answer in point form is (81/19,-3/19).

Activities
(Be Selective and just work a few)

Organize your work so it is easy to follow and write your answer in point form and indicate the quadrant. The objective is to solve for the point of intersection.

Select one to solve from problems 1-6.

1. x + 2y = 6   2. 3x + 2y = 4
3 x - 2y = 8
   x - 3y = 4

Answers: 1. (3,1.5)I   2.(3,-2.5)IV
            3. (2.2,-.6)IV

4. y = -.25x + 2 5. x + 2y - 3 = 0
   x + 4y + 2 = 0   12 - 8y = 4x

Answers: 4. No solution, ask why?
            5. Many solutions, why?

Ramp 6.4
Investigations
(Be Selective)

Geometry: Island problem

1. Three islands form a triangle with a 4th island located in the interior of the triangle. Let the four islands be represented by the triangle A, B, C, and let D be the point on the interior. The distance from A to D is 5 miles, from B to D is 8 miles, and from C to D is 3 miles.

a. Draw the picture. (Scale: 1 cm equals 1 mile)
b. The straight-line travel distance from A to B and back to A is between what two integers?
c. What is the definition of locus?

Answer: 10 < AB + AC + BC < 32

Application: Sound

2. For this activity you will need a bottle such as a glass quart milk bottle, coke bottle, or similar container, and a blower like a hair dryer.

a. Direct the air stream across the opening of the empty bottle. What do you hear? Adjust the flow of air until the sound is like a musical
b. Fill the bottle to the 1/4 mark with water and direct the air stream across the opening of the bottle. What do you hear compared to the sound in part a?
c. Fill the bottle to the 1/2 mark with water and direct the airstream across the opening of the bottle. What do you hear compared to the sounds in a and b?
d. Fill the bottle to the 3/4 mark with water and direct the air stream across the opening of the bottle. What do you hear compared to the sounds in a, b and c?
e. Fill the bottle with water and direct the air stream across the opening of the bottle. What do you hear compared to parts a, b, c and d?
f. Summarize your findings with a general statement

Thinking Problem!
A winning combination

3. Each letter in the following equation is a unique number.

$$(WIN)(WIN) = CHAMPS$$

The object is to determine the replacement digits to make two sets of digits on the left equal to CHAMPS.(Values)

Project: WIN(WIN) = CHAMPS a true statement. What are the two three-digit numbers?

  a. WIN must be larger than 316? Why?
  b. Find at least one set of values that satisfy the conditions.
Answer: a..567 ? Champ is ???

Notes

Ramp 6.5
Solving Linear Systems,
Algebraic Methods

"Mathematics is not a careful march down a well-cleared highway, but a journey into a strange wilderness, where explorers often (STUDENTS in classes) get lost."

W. S. Anglin

THE SUBSTITUTION METHOD

After completing this section, you will probably recall how to solve a two variable system by the substitution method. This method used to solve systems of equations is the substitution method. It is very similar to substitutions in athletics. Except in math the subs are equal. An example will illustrate this for you.

Given the following system, your quest is to find the values of x and y which will satisfy both equations.

$$3x-4y =-6 \text{ and } 2x + 5y = 19$$

Use the substitution method (you may pick either equation). If selecting the second equation, it can be transformed to:

$$2x = 19 - 5y, \text{ and then to}$$
$$x = 19/2 - 5y/2$$

Now we know what x is and substituting this value for x into the first equation we have:

$$3(19/2 - 5y/2) - 4y = 6$$

This is an equation you can solve since it has only one variable.

Which of the following values is y equal to? (1, 2, or 3) Answer: 3

Once you know the value of y, then substitute the value for y into x = 19/2 - 5y/2 and solve for x. Which of the following values is x equal to?

(1, 2, or 3) Answer: 2

Naturally there are other possible methods, such as you could have solved for y first and then for x. This method permits you to use your own ingenuity to solve the equations. Re-read the quotation at the beginning of this section.

Diophantus wrote one of the oldest treatises on algebra. When do you think he lived? (See comment at the end of this section.)

Activities

Solve for x and y by the substitution method. (Select one of these.)

1. $x - 2y = 11$
   $2x - 3y = 18$

2. $3x - 4y - 12 = 0$
   $5x + 2y + 6 = 0$

3. $x + 2y = 6$
   $3x - 2y = 6$

4. $3x + 2y = 4$
   $x - 2y = 8$

Answers on the x-y plane and quadrant:

1. $(3,-4)$ IV

2. $(0,-3)$ on y axis

3. ?

4. ?

Check your answers!

These are interesting
Be Selective

The following shows the point method for 3D graphing. The diagram below shows a common point which has the coordinates $(0,0,0)$. The following graph illustrates the concept.

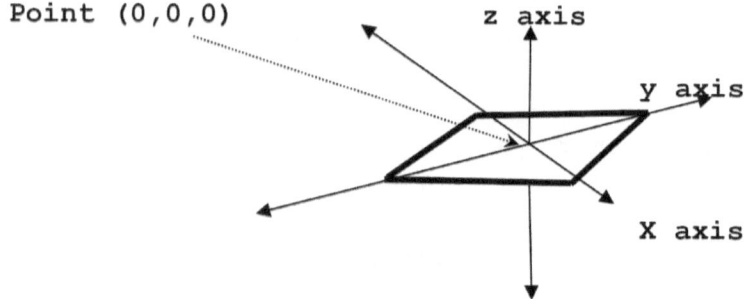

Point (0,0,0)

z axis

y axis

X axis

Can you locate the following points?

1. (3,1,5)   2. (3,2,5)

Graph your solution. To help see the location of a 3-D point, think of the 2 lines formed by the wall planes intersecting with the floor plane. The 2 floor lines are the x - y lines and the vertical line is the z line.

Interesting Investigations
Geometry: Wheels

1. Many diesel tractors have two different size wheels. The smaller wheels may make the same number of revolutions in 200 feet as the larger wheels do in 300 feet. The difference in the circumferences is 40 inches.

What is the diameter of each size wheel to the nearest tenth of an inch?
Hint: Write two equations involving r and R.

Answer: 38.2 in. and 25.8 in.

Thinking
Number theory problems

2. Diophantus died at the age of x, but from a translation of his epitaph we know the following! (Interesting problem)

>   1/6 of his years as a child
>   1/12 of his years as a teenager
>   1/7 of his years as a bachelor

Five years after his marriage his son was born. The son died four years before Diophantus did. The son lived half the years Diophantus did.

>   a. How many years did Diophantus live?
>   b. At what age did he marry?
>   c. At what age did the son die?

Answers: 84,42,33

Do a research report on: Diophantus

3. This is an example of the type of problem known as **a Diophantine Equation or problem.**

(Old problem) A farmer purchased 100 fowl for $100. Roosters cost $5, hens $1 and chicks 10 for a dollar.

    a. What is the possible maximum number of Roosters?
    b. What is the possible maximum number of hens?
    c. Why will the number of chicks always be a multiple of ten?
    d. How many of each did he buy?

Hint: Write as many equations as you can. There are 4 unknowns so you need 4 equations or conditions, which you have! They are: cost per each, and the 4th condition is the fact the answers are whole numbers

Answers are:
    a. R < 20.
    b. H < 100.
    c. 10c
    d. 9, 51, 40

Investigations
Geometry: Equation of a line

1. From geometry we know that two points will determine a line. From Descartes gift we know the equation for a line is $y = mx + b$.

    a. Given the points (–7,0) and (2,9), derive the equation for the line.
    b. b. What is the acute angle the line makes with the x-axis?

Hint: Graph the line and solve for the acute angle.

Solve the equations:

y = mx + b for m and b.

Answer: a. y = x + ? b. 45 degrees

Application:

2. *Visualization*

Below are three views of a 3-dimensional figure. Draw the 3-D view. What could it be?

Top view   Front view   Side view

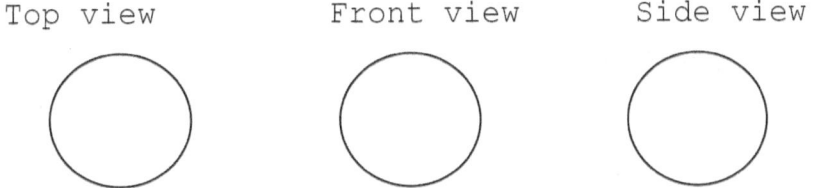

Thinking Problem: *How many socks?*

3. You are in a hurry and you have two different colored pairs of socks in the drawer. The next morning, the power is off, and you open the drawer and select socks one at a time in the dark. There are no other socks in the drawer.

a. What is the least number of socks you would have to take out of the drawer in order to insure a proper color match?

b. What is the answer for the case of three pairs in the drawer? Four pairs, five pairs, and n pairs?

c. Consider the case in "a." What is the probability of having a correct color match after selecting two socks?

d. What is the probability of having a correct color match after two socks are selected if there are three different pairs in the drawer. (Hint: Act it out!

Answers: a. 3 b. 4, 5, 6, n+1
　　　　　c. 1/3 d. 1/5

Comment: You may want to try additional questions such as: What is the probability for the third draw to provide the pair in problem "d"?

Ramp 6.6
Equations Review and Applications

The following is a listing of the postulates we used previously.

Definition 25: An equation is a statement indicating two numbers are equal

Definition 26: A formula is an equation representing a general rule such as a theorem.

Add Postulate 22: If given an equation, then you can add a number to each side of the equation.

Postulate 23: If given an equation, then you can multiply each side by the same non-zero number. (This operation includes powers and roots.)

Definition 26: A point is a set of ordered numbers written as (x,y).

Definition 27: A line, in Algebra, is defined as the set of points which satisfy the equation: $y = mx + b$.

Definition 28: Direct Variation. If two variables are so related, that for each value of y there is an x value such that, $y = kx$, then x and y vary directly. K is the constant that relates to the two variables. (This also indicates k equals y/x.)

Postulate 24: If given a set of equations, then the equations may be added (subtracted) and the result is still an equation.

Postulate 25: If given an inequation then a number can be added(subtracted) to each side of the equality and the result is an inequality of the same order.

Postulate 26: If given an inequality, then the inequality can be multiplied (divided) by a positive number and the result is an inequality of the same order.

Postulate 27: If given a set of inequalities of the same order, then the inequalities can be added and the result is an inequality of the same order.

Theorem 17: If given an inequality, then the inequality can be multiplied by a negative number and the result is an inequality of the opposite order.

If you have any questions on the above, then re-read the material. This section will provide practice in applying the above to solve applications involving equations and inequations. A suggested procedure is:

Read the problem very carefully until you fully understand it.

---

1. Define the variables by what you are trying to solve for. (Pick a variable that has meaning to the problem instead of always using x and y.)Draw a figure if needed. Write as many true equations or inequations as you can, related to fit the problem. Solve the system, if possible.
2. Answer the question or questions in the given problem. Does your answer make sense?

This is not always easy, but neither is a sport or playing a musical instrument until you work at it.

Problems
Be selective

1. A jar contains 110 coins consisting of nickels and pennies. The value of the coins is $2.50. A group problem.

Comment: This problem could be solved by trial and error, but that may take considerable time or you may be lucky. Using the steps listed above first the variables will be defined. Let N equal the number of nickels. Let P equal the number of pennies.

We know N and P will be an **integer** greater than zero. Why? We also know N will be between 0 and 50 and P will be a multiple of 5. Why?

Writing the equations:
$$N + P = 110$$
$$05N + .01P = 2.50$$
$$\text{or } 5N + P = 350$$

Solve the two equations using any one of the methods.

Answer: $35 = N$, $75 = P$

2. A student observes the wind resistance on her hand outside the car window (palm facing the wind) varies directly as the speed of a vehicle, in other words more wind pressure against the palm as the speed increases. The student estimates the pressure is 2.4 lbs. at 20 mph. Under the same conditions, what is the pressure at 60 mph?

> *Answer:* $2.4 = k20$ -->$k = .12$ and at 60 mph $P = 7.2$ lbs.

More Review PROBLEMS
(Be selective and use group discussion)

3. Given the following two equations:

$$y = -x + 6 \qquad\qquad y = x + 2$$

  a. Graph the two equations on the same set of axes.
  b. What is the measure of the acute angle each line makes with the x-axis? (Use a protractor.)
  c. What is the acute angle the two lines form at the point of intersection?
  d. Shade the region that satisfies all of the following:

   X values greater than or equal to 0.
   y values greater than or equal to 0.
   (y $\leq$ -x + 6 and y $\leq$ x + 2)

  e. Calculate the area of the indicated region. (Be able to explain your method of solution, if requested!)

Answers: a. See your graph. b. 45° c. 90° d. See graph. e. 14 sq. units

4. You have $20,000 to invest.

   a. How much would you have to invest at 7% and how much at 5% in order to have $1200 interest at the end of the year? (Guess at the answer and then work the problem.)

   b. Why do some accounts pay a higher rate of interest?

> *Answers:* a. $10,000 at each
>   b. The greater the risk, the greater the profit.

5. The general formula for a straight line is y = mx + b. What does it mean for the graph when:

   a. M in positive.
   b. M is negative.
   c. B is positive.
   d. B is negative.
   e. B is zero.
   f. M is zero.
   g. y is zero.

Answers: a. line b. line
  c. line intersect positive y-axis.
  d. line intersect negative y-axis
  e. line passes through the origin.
  f. line is parallel to x-axis.
  g. line is parallel to the y-axis.

This has been just an introduction to first degree equations and some inequalities. The second degree equations and graphics will be introduced later on. They are classified as the curves that control our lives.

Take a break!
You have earned it!

Relaxing music will help, a treat, a drink, and a joke or 2)

This week's family topic to discuss or inform the young and old as to topics such as: history of the city or state, or any selected historic items in the area.

Bridge 7: Trigonometry

A read only Bridge with a person who has had a course in Trig to assist you.

(Plan your one evening with family per week discussions.)

"TO MEASURE IS TO KNOW"
Johann Kepler

Bridge 7 *could be best understood and explained by a person who has had a course in Trig!*

Ramp 7.1
Sine and Cosine Functions
Optional

The material in this Bridge extends over a period of 4500 years and is still very useful. The first applications were solving for un-measurable distances, involving problems in construction and in navigation. Today the problems are surveying and navigation, including outer space. The playing field today is not only land and sea, but many more recent applications in the medical world. It is more important for you to have an academic foundation in mathematics today, then ever before. Keep in mind you are just touching the tip of the mathematical "iceberg

It would very informative to have a person who has had Trig read this with you to explain the definitions and operations. After completing this section, you will:

1. Understand the definitions for Sine and Cosine functions.

2. Solve a few problems involving the Sine and Cosine functions.

It was Isaac Johann Kepler, who said: **"To measure is to know."** Compare this with another quote, "**To measure the unmeasurable.**" How can you measure the un-measurable?

From that question alone, this Bridge will be interesting or at least arouse your curiosity. Can you think of a few objects that would fit into the unmeasurable list? Hint: The distance around the equator. (There was a Greek mathematician, **Eratosthenes,** who calculated the circumference of the earth about 250 B.C.) The ancient Greeks had many similar questions, like how to construct a circle with a given area. Just like you may have, this Bridge begins with some concepts, which have their origins back in the days of the Golden Age of Greek Mathematics. (If you wish, see "Golden Age" in Boyer's book, A HISTORY OF MATHEMATICS.)

Research: Trigonometry, including research as to the history of Trigonometry with regard to when, where, and why. Who was Hipparchus?

Suggested references:
> **GREAT MOMENTS IN MATHEMATICS BEFORE 1650** Eves, H.

> **THE MATHEMATICS AND PHYSICAL WORLD** Kline, M.

Or see any book on the History of Mathematics.

Ramp 7.2
The Sine Function

We have had several new terms in this Bridge. Can you identify the terms? The first word needing explaining or defining is used in the title for this Bridge, Trigonometry. What do you see in the word, trigonometry? Your first answer is probably "metry", which is associated with measure. The second word you would pick out is "tri", which is associated with three as in triangle. Most students then ask where the middle part "gono" fits in. The answer is the Greeks actually used the word trigon or trigonon, which means, triangular in shape. **Putting these two together we have triangle measure or trigonometry. (What is etymology?)** The meaning of the terms SINE and COSINE will be given at the appropriate point.

How did the Greeks develop this branch of mathematics and why? The why may be easier to explain then the how. We have weather forecasters today and you probably hear their predictions every morning on radio or TV. In the ancient days these forecasters predicted the times for planting, for harvesting, and the days for what we call holidays. We also have land surveyors who are usually checking the boundaries for a lot, or determining boundaries or key points for a freeway or toll road. The ancient surveyors determined the boundaries after a flood to determine which land belonged to the king, etc. One of their tools was a piece of rope marked off in 3-4-5 segments. Can you figure out why? (Hint: Pythagorean Theorem) **The surveyors of the early times were also called rope stretchers.** Their tool was as indicated above (the 3 by 4 by 5 rope) which when stretched at the vertices of 3-4-5 a right triangle is formed. It was a convenient way to determine a corner for a lot. You may wish to make this tool and demonstrate the method to your friends. Today of course, modern surveying instruments are used.

Comment: Discuss surveying with a local surveyor or invite him to one of your group meetings and discuss his tools and methods. Below is the drawing of several right triangles.

Figure 1

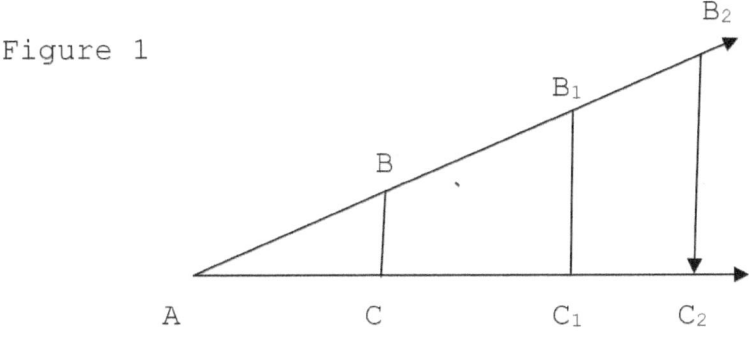

How are the above triangles related? The answer is, as you recall from geometry, they are similar. Two right triangles are similar if two angles of each triangle are equal, or if the condition SAS = SAS exist.

Postulate 28:
The Triangle Similarity Postulate

How many triangles are there in Figure 1 above? **From the fact they are similar, we also know that the corresponding sides are in equal ratios.**

Hint: Map the triangles above to enable you to pick out the corresponding sides. Therefore, we can write:

$$\frac{BC}{B_1C_1} = \frac{AB}{AB_1} = \frac{AC}{AC_1}$$

Write two more sets of ratios.

By using algebra, these equations can be written as ratios. Another way to show the ratios is to map the triangles. Here is an example:

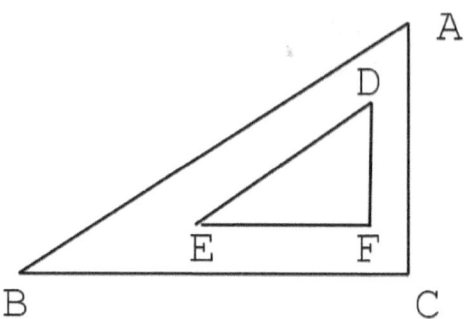

What does this really say (converting the statement to if-then form)? If two right triangles are similar, then the ratio of the corresponding sides is a constant and the corresponding angles are equal. The Greeks named this ratio the SINE of the angle.

Sin A = C/AB.

Comment: The term sine they think came from the Hindus with reference to the chord of a hunter's bow. This term will be further explained in another section of this Bridge.

Activity

1. Draw a large 30-60 degree right triangle on your paper and using your protractor draw perpendiculars, so you have at least several right triangles, similar to the following figure. Angles C and C1 are 90 degrees.

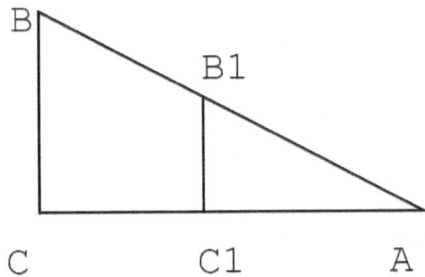

If m∠A = 30 degrees, can you calculate the angle of B and B1?(Use your protractor to measure the angle.)

Definition 29: The sine of the acute angle is the length of the side opposite the angle divided by the length of the measure of the hypotenuse.

Sin A = a/c
Sin B = b/c

Fortunately, these ratios have been worked out for you and are in your calculator. Find the sine button on your calculator. In order to find the sine of the acute angle your calculator must be in the degree mode. (If you don't have your book of instructions that came with your calculator, then ask a math teacher to help you tell when your calculator is in degree mode.)

Example: What is the sine of 37°?

Steps:	Display
1. Turn calculator on.	0
2. Press key 3 then	7. 37
3. Press sin key.	.6018
(4 decimal places)	

Comment: If you did not get that answer, check the instruction book that came with your calculator, or ask another person who did get the answer.

The activities for finding the trig values for the angle and the reverse activity need not be completed once the use of the calculator is routine. This will save you a lot of time!

Activity
(Use your calculator)
Work a few of these until you
feel that you understand.

1.What is the sine of the following angles? (Angles are in degrees.)

a.	10	b.	13	c.	30		
d.	45	e.	55	f.	60		
g.	75	h.	82	i.	85	j.	90

k. What do you observe as to the value of the sine function as the angle increases?

Answers:

a.	.173648	b.	.224951	c.	.500000
d.	.707107	e.	.819152	f.	.866025
g.	.965926	h.	.990268		
i.	?	j.	?		

2.This time the problem will be reversed. Using your calculator, record the measure of the angles whose sines are listed below. (Answer to the nearest degree.)

Comment: You may need to check the instruction book for help with this inverse operation. (Work a few until you feel you understand.)

    a. Sin A = .42266
    b. sin A = .60185
    c. sin A = .95630
    d. sin A = .65560
    e. Sin A = .45399

Answers: a.25^0 b.37^0 c.73^0
               d.41^0 e.27^0

Ramp 7.3
The Cosine Function
(Pronounce co-sign)

For this function follow the same method as you did for the Sine. In the figure below the method used for labeling angles and sides is capital letters for vertices and small letters for the sides. The small "a" represents the side opposite the vertex A, like-wise the small "b" is opposite vertex B. This is nothing new to you since this method was used consistently in your geometry course and possibly other previous courses. See the figure below.

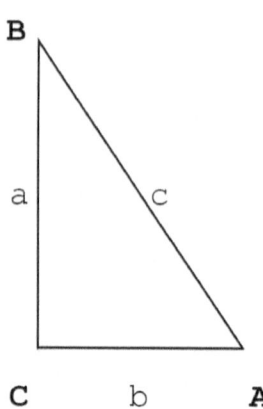

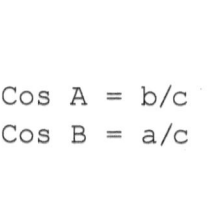

Cos A = b/c
Cos B = a/c

The **COSINE** (abbreviated COS) of an acute angle in a right triangle is the ratio of the measure of the adjacent side divided by the measure of the hypotenuse.

Definition 30: The COSINE of an acute angle in a right triangle is the ratio of the length of the adjacent side divided by the length of the hypotenuse.

As a challenge you may attempt to justify the following.

Given: Right triangle ABC is similar to triangle DEF with angles C and F right angles. m∠A = m∠D.

Map the two triangles.
Justify: The ratio cosB = CosE or that a/c = d/f. In Figure 2, the cosine of angles A and B are listed below.
Cos A = b/c, Cos B = a/c

Example: What is the cos 43°?

Steps	**Display**
1. Turn calculator on.	0
2. Press buttons 4 then 3.	43
3. Press cos button.	.7314(4 places)

Comment: If you did not get that answer, ask another person or check your calculator instruction book.

Activity
Be Selective

1.What is the cosine value for the following angles? (The measure of the angles are in degrees.)

a.17   b.26   c.30   d.4
e.56   f.60   g.72   h.87   i.89
j. As the measure of the angle increases what does the value of the cosine approach?

Answers:
a. .956304   b. .898794
c. .866025   d. .707106
e. .559193.   f. .5
g. .309017   h. .052336
i. .001745
j. The value approaches 0.

2. What is the degree measure of the angle for the following?

a. Cos A = .29237
b. Cos B = .48481
c. Cos A = .91355
d. Cos A = .54464
e. Cos B = .97437
f. Cos A = .08716
Answers are in degrees rounded to the nearest degree.

a. 73 b. 61 c. 24 d. 57
e. 13 f. 5

Now you are possibly beginning to think: "What are the sine or cos functions used for?" Two examples and then some practice problems.

In the following exercise, an engineer is given the problem to determine the length for a future water line indicated by a, using the given information indicated.

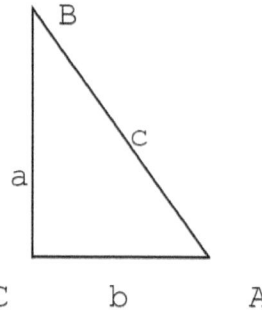

m∠A = 40°, side a = ?,
hypotenuse = 110 ft.

Solution: The side the student needs to solve for is opposite the given angle. This means you need to use the sine function, since the side opposite is the side in question and the hypotenuse is known. (See definition of sine.)

Then writing the equation:

Sin 40°= a/110

Substituting the value of sine 40° from your calculator and then solving for the unknown results, side **a** is equal to 70.71 feet rounded off.

$$.6427888 = a/110$$
Solving for **a** by algebra:
$$110(.642788) = a$$
$$a = 70.706637 \text{ ft.}$$

Do you think this is a good answer? Why all the decimal places? We will agree to always give the answer to fit the original data! What do you think the 110 ft. was measured with? Let's assume a measuring device like a tape measure and the measurement is to the nearest foot, since the 110 doesn't have inches or a decimal fraction indicated. Therefore, we will give our answer to the nearest foot or 71 ft.

Another example: A ramp is 12 ft long and 5 ft high as indicated.

What is the angle of the incline?(angle A)?

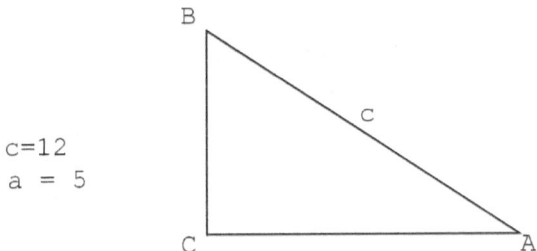

c=12
a = 5

Dividing 5/12 with your calculator the display reads .41667 (rounded).

Now we need to know the angle that has a cos equal to .41667.

Press inverse or second function button on your calculator, then the cos button and the display should read 65.3756 (depending on your type of calculator, see instructions) or rounded to 65.38 degrees or 65 degrees and 23 minutes.

Comment: The method you use in rounding answers is important. The answers should make sense considering the data in the problem. The answer should have at least the same degree of accuracy as the values in the problem.

In this section you have two definitions to understand!

The above should be recorded in your notebook with examples, and memorized.

In working the following activity, keep in mind the trigonometric functions at this point only work with right triangles. In other words, to use the sin or the cos functions you must have a right triangle. If the right triangle isn't given you must create one in the figure by your ingenuity. Draw a figure for each problem.

Activity
More practice if needed
Be Selective! Work a few.

1. Using your calculator, write the value for the sine of each of the following angles rounded to 5 decimal places unless the answer is exact. (Angle measure is in degrees.) You no doubt need to work a few of these.   Answers:

	Angle (degrees)	Sine
a.	5	.08716
b.	10	.17365
c.	15	.25882
e.	20	?
f.	25	?

continue if need to.

2. Which angle or angles in problem 1 have exact values for the Sine? Cosine? Tangent?

Answer: Try angles 30,45,60

3. What number does the value of the sine appear to approach as the angle gets larger in problem 1? As the angle gets smaller?

Answer: 1, 0

4. Using your calculator write the value of the cosine for the following angles. Round your answer to 5 decimal places, unless the answer is exact. (Work a few and especially the **30,45,60, and 90** angles.)

Angle (degrees)	Cos
a. 5	.99619
b. 10	.98481
c. 15	.96593
d. 20	?
e. 25	?
f. 30	?
g. 45	?
h. 60	?
try 90	

5. Which angle or angles in problem 4 have exact values for the cosine?

Answer: 60 degrees

Hint: The cosine of an angle is the same as the sine of the complement.

6. Convert the following to degrees and minutes.

   a. 16.4 degrees is equal to 16 degrees and ? minutes.
   Hint: .4 of 60 equals ?

   b. 61.7 degrees is equal to 61 degrees and ? minutes.

   c. 57.5 degrees is equal to 57 degrees and ? minutes.

Answers: a. 24′ b. 42′ c. 30′

7.In the famous 3-4-5 right triangle, what is the measure of the smallest angle to the nearest degree?

Answer: 37 degrees

8. a. Do you think the sin of 25 degrees is 1/2 the sin of 50 degrees? Justify your answer.
   b. The sine of 25 degrees is 1/2 the sine of what angle?

Answer: a.Your explanation
b. 57.7 degrees

Ramp 7.4
Justification for Pi
(A Must For All)

This is very interesting! Read it very carefully! The original problem was created when a person asked the engineer, 2500 B.C., to construct a circle with a circumference of 100 feet.

A justification:(This should be very interesting solving for the value of Pi!)

The Value of Pi!

Pi is defined as the ratio of the circumference to the diameter or 2 radii. (The perimeter of the inscribed regular polygon approaches the circumference of the circle as the number of sides of the polygon increases!)

Given: Chord AB with midpoint D. OD is a perpendicular bisector of **Chord** A D B.

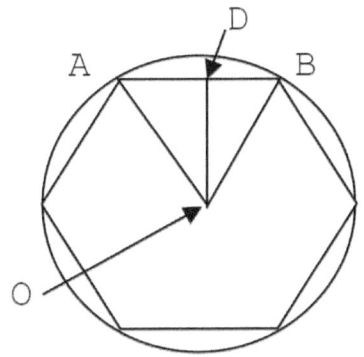

Solution:

Sin of angle AOD is AD/AO and as the number of sides of the inscribed polygon increases to n sides then: Angle AOD is 90/2n as the number of sides increase. Side AD is ½ the side of AB of the polygon. Sin(90)/n or AD will approach 1. Recall: The definition of PI.(The perimeter P is 2n(AD) and approaches 2KR, where K is a constant which Euler gave it the name Pi.)The number of sides of the polygon increases and the perimeter approaches the circumference of the circle. Select n for AD, P equals 2n(R){(sin(180/n), R = 1}

Pi is C/D and the perimeter is very close to the circumference of the circle when n is very large and we can write:

Perimeter/2R approaches K,
= (sin(90/n)(2n). is the limit of
(sin(90/n)(2n)as n increases.

Values: (round the value to 5 Decimal places)

n	
16	3.13655
32	3.14033
64	?
128	?

Calculate the value of Pi using the above formula for the following values of N. Record your answers, as the number of sides increase:

      N = 256, 516, 1032.
      Write your comments for the
      value of Pi.

See Beckmann's interesting book, HISTORY OF Pi. (Especially the Indiana Case.)

Euler named the k constant Pi which is the first letter in the Greek word for perimeter.

Thinking problem:
The Schedule

It is Sunday and you are planning your schedule for the week, and hopefully you will be able to make only one trip to town. You live 15 miles from town and are only free in the afternoons. You have to visit the dentist, buy bakery goods, and select sweet corn at the Farmer's Market. The dentist's office is open only on Tuesday, Wednesday, and Friday. The

Bakery is closed on Wednesday and Saturday afternoon. The Farmer's Market is not open on Tuesday and Thursday. When should you schedule your trip to town? Justify your answer.

> *Answer:* Friday afternoon is the best time to go.

Ramp 7.5
Tangent Function
(Optional)

"Hipparchus of Nicaea, (180–125 B.C.E.) compiled the first trigonometric table."

Boyer, C.B.

The TANGENT FUNCTION

(Work this function the same way as you did the Sine and Cosine functions.)

After completing this Bridge you will:

1. Understand the Tangent function.
2. Be able to solve problems using the Tangent function.

In section one of this Bridge, you worked with the sine and cosine functions or ratios. The condition was that you had to have a right triangle. When you hear the phrase right triangle one important theorem comes to your mind or should, and it is the Pythagorean Theorem. We are now studying the trigonometry of the right triangle. The sine ratio involves the opposite side and the hypotenuse. The cosine ratio involves the adjacent side and the hypotenuse. You can probably guess what the tangent ratio involves.

We will start this time with a definition.

> **Definition 31:** The Tangent of an acute angle in a right triangle is the ratio of the length of the side opposite the angle divided by the length of the side adjacent to the angle (not the hypotenuse).

Using the same lettering, caps for points and small letters for segments.

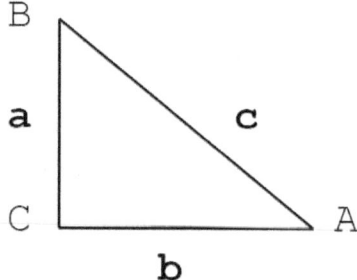

Tan A = a/b or using angle B the Tan B = b/a read as the tangent of angle A is the ratio of side a to side b.

Activity
Be selective

Since this activity is similar to the Sin and Cos activities you may not need to work all these practice activities.

1. What is the tangent value for the following acute angles? Use your calculator. Work a few to learn the use of the calculator.

Angle measure in degrees.

Angle degrees	Tangent Values
a. 10	.176326
b. 21	.383864
c. 30	?
d. 45	?
e. 47	1.072369
f. 60	?
g. 79	5.144554
h. 82	7.115370
i. 87	19.0811
j. 90	?

Which angle has a tangent = to 1?

2. The value of the tangent is given and you are to read the measure of the angle from your calculator. The first problem is worked for you. Be sure to get the same answer?

Tan A = .1234, then the measure of angle A is? (Ask your friend or read the instruction book for your calculator.)

Turn calculator on and enter.1234. Push the "inv", or "sec F" key, then the Tan key and the display should read 7.03 rounded to the nearest hundredth. Some calculators have "Tan⁻¹" key which gives the answer directly. What are the angles for the following?

a. .1405	b. .2679	c. .4663
d. .7265	e. 1	f. 1.80072
g. 5.1446	h. 11.4301	

Answers: a. 8 b. 15 c. 25
d. 36 e. 45 f.61
g.79 h.85

All values rounded off to degrees.
How is this tangent function used?

3. This example involves road construction. In the figure below a new road is being planned in order to move traffic more efficiently from A to B. The old road is now A to C and then C to B. Angle ACB is a right angle and B cannot be seen from A. What angle should the new road make with line segment AC?

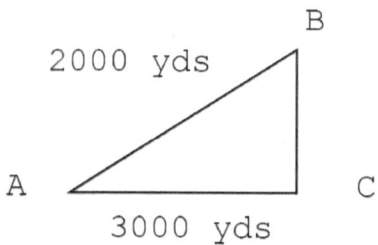

B
2000 yds
A
3000 yds
C

Explanation:

The measure of angle CAB is required. The known sides are opposite and adjacent with regard to angle A. Therefore, the tangent ratio will be used (see definition). The equation is Tan A = 2000/3000.

The answer is 33.7 degrees or 33° and 42 seconds. The following Home Activity will involve the three functions, so in some of the problems you will have to decide which ratio or function you will use.

The three definitions or functions are:

Definition 29: The SINE of angle A in a right triangle is the ratio of the length of the side opposite the angle divided by the length of the hypotenuse.

Definition 30: The COSINE of an acute angle in a right triangle is the ratio of the length of the adjacent side divided by the length of the hypotenuse.

Definition 31: The Tangent of an acute angle in a right triangle is the ratio of the length of the side opposite the angle divided by the length of the adjacent side to the angle.

Draw the right triangle and label it with the given data.

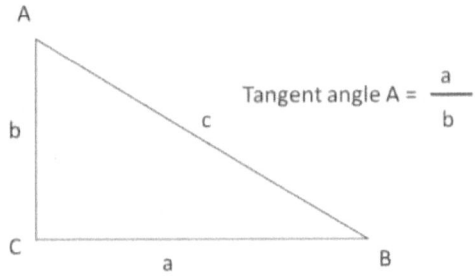

a. Indicate on the figure the angle or side you need to find (hint label the sides of the triangle a,b,c and the opposite angles A,B,C).
b. Give the lengths of the sides
c. Write a trigonometric equation involving the data.
d. Solve the equation.
e. Answer the question.
f. Does the answer make sense?

Activity
Be Selective
Use a calculator

1. Determine the Tan value for the following angles. The angles are in degrees and you are to use your calculator. Round the answers to five decimal places.

(Work a few until you feel competent!)

Angle	Tangent values
a. 5	.08749
b. 10	.17633
c. 15	.26795
d. 20	.36397
e. 25	.46631
f. **30**	.57735
g. **45**	1.00000
h. 55	1.42815
i. **60**	1.73205
k. 90	?

2. What do you observe about the Tan values as the measure of the angle increases?

Answer: Increases

3. What is the measure of an angle in degrees that has a Tan equal to 1?

Answer: ?

4. A student brought the following problem to class and asked the class to help solve for the length of ZY, the approximate width of the river. In other words, solve for the "unmeasurable." A surveyor measured angle XZY as 43 degrees with a transit, and segment XY was 105 feet.

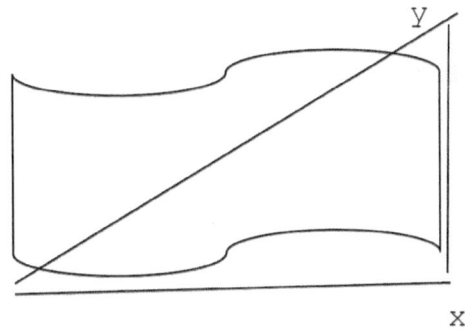

a. One person said the problem could be solved by using the Tan function. Was the student correct?
Write Tan 43 = ?/?.

b. What is the length of YZ?
c. Did you measure the unmeasurable?

Answers: a. Yes. Tan 43 =.9325
b. 98 ft.   c. yes

5. What are the angles in a 5-12-13 right triangle (to nearest degree)?

Answer: 23, 67, 90 degrees

6. One student excitedly proclaimed that she discovered the following, which appears to be valid for all angles!

$$\text{Tan A)}^2 + 1 = 1/(\cos A)^2$$

a. Check this by substituting values for A, where A is in degrees.
b. Do you think it is always true for acute angles?

c. Can you justify that it is always a valid equation for acute angles?
Hint: Let Tan A = a/b and
Cos A = b/c in the right triangle

Answer:$(b/a)^2 + 1 = 1/(b/c)^2$
This simplifies to $a^2 + b^2 = c^2$
which is valid for any right triangle.

7. A surveyor measures a right triangular lot and reports to his employer that the sides are 40 meters by 50 meters by 90 meters. If you were the employer, what would you say to the surveyor?

Answer: Measure again.
Can't be atriangle!

8. **Research:** How Eratosthenes derived the circumference of the earth is explained in Hogben's book ***THE WONDERFUL WORLD OF MATHEMATICS*** (see page 36. Very informative reading!).

9. Your Community

a. What is the latitude of your community?
b. How far are you from the North Pole?
c. What is the circumference of the circle of latitude for your community?
d. What is the longitude of your community?

Notes and Comments:

Ramp 7.6
Trigonometric values for any Angle

**"The advance and the perfecting of
mathematics is closely joined**

―――

to the prosperity of a nation."

Napoleon

Sin A = ?   Cos A = ?   Tan A = ?

Up to now the Trig used only right triangles and you probably asked the question, what if the triangle isn't a right triangle? The answer is coming.

After completing this section, you will understand:

1. Definitions of the Sin, Cos, or Tan of any angle.
2. With your calculator, be able to determine the value, if there is one, for the Sin, Cos, or Tan of any angle.

In the last section, the definitions of the three basic trigonometric ratios were stated for the acute angle of the right triangle, and you solved a few problems involving these functions. You notice that sometimes the term ratio has been used, and at other times, the word function has been used. Both terms are correct, but they do have different meanings. You, no doubt, also used the term function in connection with algebra, but like any new word, the meaning only becomes understood by continued use.

The sin of an angle is a ratio, and it is also a function. Why is it a function? It is a function, because for each degree (x) value there is only one y value. This idea is also illustrated in point form, which means for each x there is only one y, where a point is defined as a set of ordered pairs (x,y). This is why the term function is used.

Eventually, the meaning will be understood, and you may even begin to use the term in expressing your ideas.

The trig ratios are functions of the angles, even if the angles are positive, negative. than 90 degrees. The reason the number 90 is used is that in all three definitions (Sin, Cos, Tan) the angles were limited to acute angles. Don't you wonder how the mathematicians created or devised the system to handle all types of triangles, obtuse, acute, and eventually any size angle?

Trig. Functions for any angle
The meaning of Sin, Cos, or Tan for any angle.

The geometry you have been studying is mostly from the ancient Greeks, 600 B.C.E. to 100 B.C.E.

Now we will jump to the early 1600's C.E., approximately 1800 years later. The world of ideas is constantly changing today as it was in the 1600's. (Can you think of a few new inventions in your lifetime? Ask your friends the same question.) A philosopher-mathematician by the name of Rene Descartes (pronounced Day-Cart) is given credit for inventing or adapting the geometry of Euclid to the xy plane.

Research: The interesting life of Descartes including:

    a. His dreams,
    b. His inventions,
    c. His army life,
    d. His school days,
    e. The Swedish Queen in his life,
    f. Where he died.

History

Suggested Reference:
E.T.Bell's, *MEN OF MATHEMATICS*,
Simon and Schuster 1965, chapter 3
(This makes for an interesting bit of history.)

The problem of how to determine the Sin of an angle greater than 90 degrees was a problem for mathematicians for many years. You can tell from the spread in years between Euclid and Descartes.

How did Descartes explain it?

Activity: Descarte's method

1.Take a sheet of graph paper and draw the xy axes as you did in algebra class.

2. On the axes draw an angle with the vertex at the origin (0,0), and one ray on the positive side of the x-axis. The other ray is rotated in a counter clockwise direction or in the first quadrant direction so it is in the second quadrant. See the figure below.

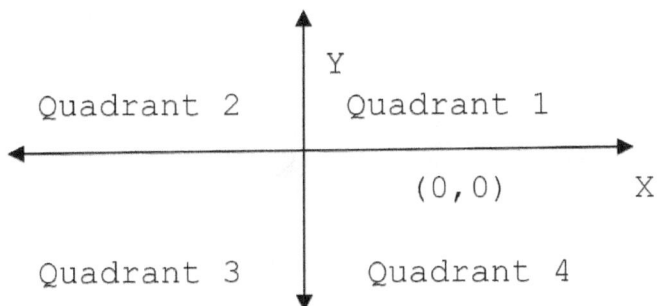

3. On your xy plane, draw the following angles using the method described in number 2. Use your protractor. All the angles are measured in degrees.

   a. 30, 60, 120, 150, 210,
      225, 240, 300, 345, 360,
   b. On each ray, identify a point five centimeters from the origin (0,0).
      From the point on each ray,
      draw a perpendicular to the x-axis.

What Descartes did next was to generalize the definitions for the Sin, Cos, and Tan for the acute angles. He observed the triangles were similar to the corresponding acute right triangle in the first Quadrant. Therefore, the ratios are the same except for the sign! Then he adapted the definitions to fit the general angle. The definitions are stated below. Perhaps, you had better read the above again.

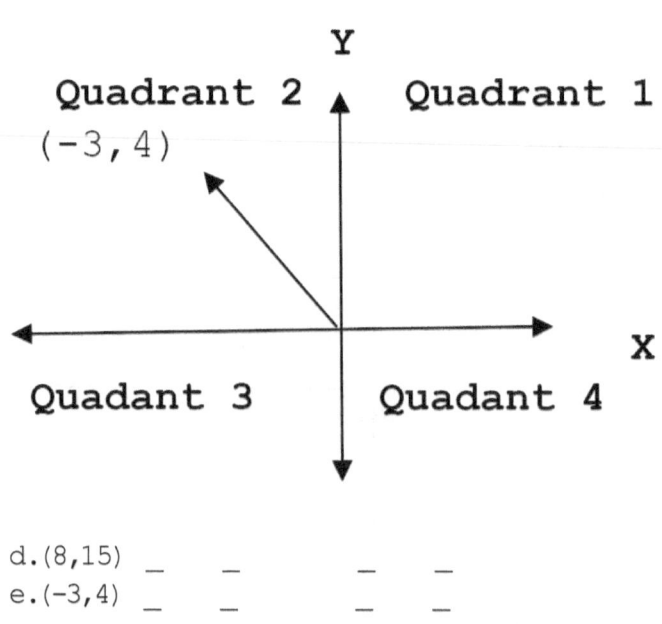

Y

Quadrant 2
(-3,4)

Quadrant 1

Quadant 3

Quadant 4

X

   d. (8,15) _ _   _ _
   e. (-3,4) _ _   _ _

Answers:

b.	13	12/13	5/13	12/5
c.	10	8/10	6/10	8/6
d.	17	15/17	8/15	15/8
e.	5	4/5	–3/5	–4

Use your calculator to complete the following Record the answers rounded to 4 decimals places.) Draw each angle with vertex at the origin as shown above. The first row is completed except for the sketch. All angle measurements are in degrees for now.

Be selective!

Angle(degrees)	Sin	Cos	Tan
a. 40	.5878	.8090	.7265
b. 70	?	?	?
c. 85	?	?	?
d. 95	?	?	?

One student made the following observation:

If the Angle is:

**Case 1: *0 < A < 90*,
then the three trig functions
are all positive.**

**Case 2: *90 < A < 180*,
then Sin is +,
Cos is –, Tan is –.**

**Case 3: *180 < A < 270*,
then Sin is –,
Cos is –, Tan is +.**

**Case 4: *270 < A < 360*,
then Sin is –,
Cos is +, Tan is –.**

The objective of this section is to enable you, with the help of your calculator, to find the value of the trig functions for any positive or negative angle. The ratios for **negative** angles are evaluated the same way as you evaluated them for positive angles, but the terminal ray of the angle is rotated in a clockwise direction. The initial ray is always on the positive x-axis and the vertex is at the origin or the point is (0,0).

Example:

What is the Sin, cos, and Tan for an angle of -65 degrees. The easy way to find the values is to punch into your calculator the -65 and then hit the Sin button. Sketch the angle.

   1.Draw the figure to show a
     -65 degree angle

   2.Put in -65(degree mode)

   3.Click the Sin button
     Sin -65 = -.8526(4 places)

   4.Same procedure for the other
     Functions:
          Cos -65 =.4226
        Tan -65 = - 2.1445

Comment: **This section has re-defined the Sin, Cos, and Tangent functions to work for any size angle. The up-dated definitions are stated as follows.**

Definition 32: If the vertex of the angle is at the origin with one ray on the positive x-axis and the point (x,y)is on the other ray, then:

$$Sin\ A = y/R \qquad Cos\ A = x/R \qquad Tan\ A = y/x$$

Conditions:
a. Where R is the distance the point is from (0,0). R is always positive, since distance R is the positive square root of $(x^2 + y^2)$. The vertex is always at the origin and the positive x-axis is one ray for the angle.
b. Positive angles are rotated counter clockwise.
c. Negative angles are rotated clockwise.

Many students find the following information useful.

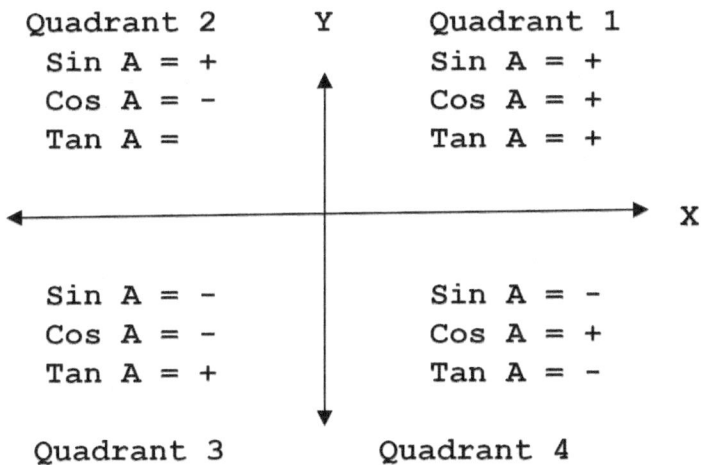

Quadrant 2   Y   Quadrant 1
Sin A = +   Sin A = +
Cos A = -   Cos A = +
Tan A =   Tan A = +

Sin A = -   Sin A = -
Cos A = -   Cos A = +
Tan A = +   Tan A = -

Quadrant 3   Quadrant 4

This tells the sign of the Trig functions in each quadrant. The vertex of the angle must be at the origin (0,0), and one ray on the positive x-axis.

Activity
Be Selective

1. The point is a (12,16)
   a. Solution:
$$R = \sqrt{144 + 256}$$
$$R = \sqrt{400} =$$

2. Complete a few of the following.

Angle(degrees)		Sin	Sketch Cos	Tan
a.	25	.4226	.9063	.4663
b.	-25	?	?	?

Answers: -.4226 .9063 -.4663

| c. | 30 | ? | ? | ? |

Answer: .5 .8660 .5774

| d. | -30 | ? | ? | ? |

Comment: The values of the trig functions differ from the values in first quadrant only by sign, or they have the same absolute value.

APPLICATION
Investigations
Be selective

Application: Speed of rotation

1. The earth revolved once in 24 hours. If the circumference is approximately 24000 miles then how many miles per hour is a point on the equator revolving, approximately?

Thinking: Digital possibilities

2. A counting number consists of three digits, 9, 5, and x in that order. If the number is reversed and then subtracted from the original number, the answer will have the same digits, but a different order.

---

a. How many possibilities are there for x?
b. How many possibilities are there for x, if repetition is not permitted?
c. How many different 3 digit numbers (All digits being different.) Can you write using 9, 5, x?
d. What is the number that satisfies the condition in the statement of the problem?

Notes and Comments

Ramp 7.7
More Trigonometry

"The heart of the mathematical experience is of course, mathematics itself."
Davis, P. and Hersh, R.

THE MATHEMATICAL EXPERIENCE
Trig For NON-RIGHT TRIANGLES

None of this, you may have had, even
if you have had a course in Trig.

After completing this Bridge you will better:

1. Understand the Sine Theorem,
2. Understand the Cosine Theorem.

In the last section you learned how to find the Sin, Cos, or Tan of any angle, even negative angles. In this section, you will learn how to use these functions to solve problems involving triangles, with no limitation as to type of triangle. In other words, the triangle doesn't have to be a right triangle. Many of the problems in the real world involve right triangles but many do not, such as land surveying, road construction, and navigation on the land, sea, and in air or space. It also plays a vital role of power factor connected with electrical energy transmission as well as many medical related applications.

The solutions to **non-right triangle** problems involve two theorems, the Sin Theorem and the Cos Theorem. These are useful theorems since they work for right triangles as well non-right triangles.

If you know these two theorems and can apply them, you can solve many applications involved in architecture, navigation, medical, military, surveying, etc.

The Sine Theorem

The name of the inventor of the Sine theorem has been lost in history so a report can't be recommended pertaining to that person and his or her life. The following problem is one that the ancient mathematician tried to solve. In the figure below, let's assume the point S is a ship, and the points A and B are on the shore 10 kilometers apart Angles A and B are 40 and 30 degree, respectively

This could be a military or a rescue problem where S is the ship and the defensive guns or rescue teams are at A and B. Distances AS and BS are required.

The "guns" may have been rock throwers or catapults in ancient times and rockets today. Now that you are hopefully curious as to how the problem is solved, the method will be explained. First, the general solution will be derived. Follow these steps, ask questions where explanations are needed.

Step 1. Draw a triangle and label the figure.

> **Comment:** The **general case** does not use values, but only letters or variables in this case A, B, and C for the angles, a, b, and c for the sides.

Step 2. Convert the drawing to right triangles by drawing the perpendicular or plumb line CD. This is done so you can use the theorems for the right triangle from the previous sections. Triangles ADC and BDC are right triangles.

Step 3. Notice AC is labeled b, BC as a, and CD as h to make communication easier. Do you agree with the following?

   a. Sine A = h/b
   b. Sine B = h/a

Step 4. If Sine A = h/b, then h = b(Sin A).
   If Sine B = h/a, then h = a(Sin B)

Step 5. Since h = h, and by substituting from step 4 you have:

$$h = h$$
$$b(\text{Sin } A) = a(\text{Sin } B)$$
$$\text{Dividing by ab gives:}$$
$$\frac{\text{Sin } A}{a} = \frac{\text{Sin } B}{b}$$

Step 6. If the perpendicular had been from A to BC the proportion would have involved Sin C over c. Try it!

Consequently, the following theorem can be stated.

Sine Theorem 19:
If given a non-right
triangle, labeled ABC, then:
$$\frac{\text{Sin A}}{a} = \frac{\text{Sin B}}{b} = \frac{\text{Sin C}}{c}$$

You can easily see why it is called the Sine Theorem? It should be easy to remember! Now back to our problem.

Draw the figure and label it to fit the theorem. We know that AB is 10 kilometers, angle SAB is 40 degrees and angle SBA is 30 degrees.

What is the measure of angle ASB?

Answer: 110 degrees.

Write the equations using the Sin theorem:

$$\frac{\text{Sin A}}{a} = \frac{\text{Sin B}}{b} = \frac{\text{Sin S}}{s}$$

$$\frac{\text{Sin 40}}{a} = \frac{\text{Sin 30}}{b} = \frac{\text{Sin 110}}{10}$$

$$\frac{\text{Sin 110}}{10} = \frac{\text{Sin 40}}{a} = \frac{\text{Sin 30}}{b}$$

therefore, $a = \dfrac{(Sin\ 40)10}{Sin\ 110} = \dfrac{.6428(10)}{.9397}$

BC or a equals 6.84 kilometers or approximately 4.25 miles. In like manner the distance b can be determined using

$$\frac{\text{Sin B}}{B} = \frac{\text{Sin A}}{10}$$

Substituting:

$$\frac{\text{Sin } 30}{} = \frac{\text{Sin } 150}{}$$

$$\frac{\text{Sin } 30}{1000} = \frac{\text{Sin } 110}{b}$$

and solving for therefore,

$$\text{AC or b} = \frac{(\text{Sin } 30)10}{5.32 \text{ km.}} = \frac{.5(10)}{} =$$

$$\text{Sin } 110 = .9397$$

This means the distance from A to C is 5.32 km or approximately 3.31 miles. The final answers are rounded.

You must admit these ancient mathematicians were very clever. This also makes it evident why schools teach algebra before geometry.

Activity

1. An engineer for a road construction company is given the problem of calculating the length of the approximate new section BA.(Indicated on the map below as AB) It is known that ∟A is 42°, ∟C is ∟56°, and BC is 3 miles. (Not drawn to scale) Use the Sin theorem.

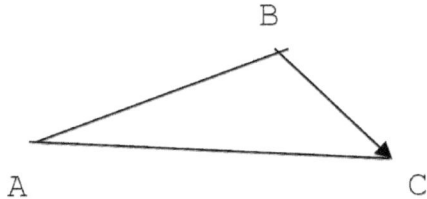

Answer: AB = 3.7 mi.

This time a city surveyor came back to the office with the following measurements for a city park. The mayor needs to know the measure of the angle at C.

SUGGESTION: Make a drawing to scale.

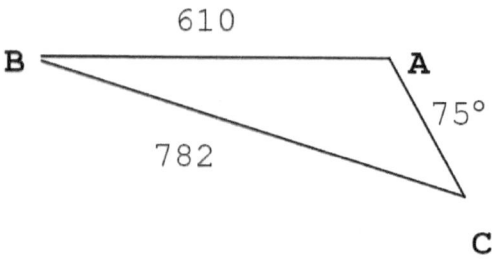

Is the following equation correct?
$$\frac{Sin\ C}{610} = \frac{Sin\ 75}{782}$$

Use your calculator and solve for angle C.

Answer: 48.9°

Can you solve for b and also the other side? If your answer is yes, then solve for angle B.

Answer: B is 56.1 degrees and The side is 672 (assume feet).

Do you see why the quote was selected at the beginning of this Bridge and why the author thought it was appropriate?

"To Measure The Unmeasurable."
Johann Kepler

The following practice problems can be solved using the Sin theorem. Round the answer or answers one which makes sense to you according to the data in the problem.

The following procedure or steps will help you solve the procedure:

a. Draw the figure.
b. Label the figure.
c. Identify the unknown.
d. Write the equations.
e. Solve for the unknown.
f. Answer the question.

Ramp 7.8
The Cosine Theorem

What if the triangle isn't a right triangle?

One day after thinking about the Sin Theorem, a student asked the professor; "How do you solve a non-right triangle problem if all that is given are the three sides, or two sides and the included angle?" Very good question, since the Sin Theorem will not handle this case due to the fact that in order to use the Sin

Theorem you must an angle and the opposite side plus one more fact -- a side or an angle. The professor said: "use the Cosine

Theorem!" The following problem will help you understand the Cosine Theorem.

Given: Triangle ABC with AC = 5, BC = 7, and AB = 9

Question: What is the measure of the angles to the nearest degree? Draw the figure.

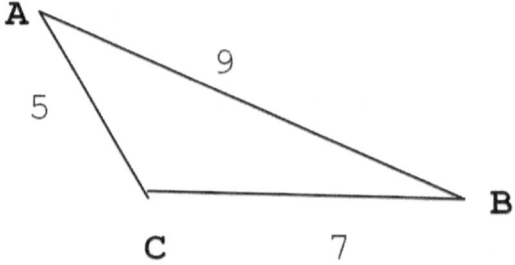

In order to solve this problem, the professor showed the student a more general problem and informed the student, that if the general problem is solved, then we just need to adapt the specific problem to the general formula and substitute the values into the formula.

General case: Draw triangle ABC on the Cartesian plane or on the x-y axes as shown with sides labeled a, b, and c. Using the drawing above adapt the following information to the figure.

Step 1: In the figure below, from A, draw a perpendicular to BC, extended, and label the additional segments h and x as shown.

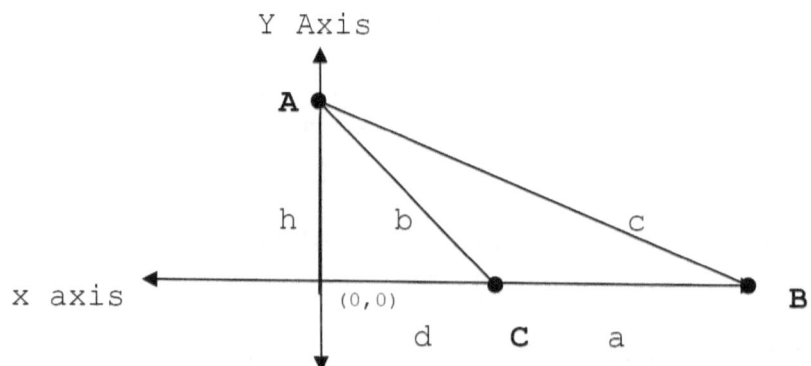

Step 2. Now using algebra and the Pythagorean Theorem, check the following:

$$c^2 = h^2 + (d + a)^2$$
by algebra: $c^2 = h^2 + d^2 + 2ad + a^2$

commuting the terms:
$$c^2 = a^2 + h^2 + 2ad + d^2$$
but $b^2 = h^2 + d^2$ or
$h^2 = b^2 - d^2$ (see figure)

Substituting $b^2 - d^2$ for h^2 in the equation above.
$$c^2 = a^2 + b^2 + 2ad$$

Substituting: $d = b(\text{Cos ACD})$
since $\text{Cos ACD} = d/b$.

This results in:
$$c^2 = a^2 + b^2 + 2ab(\text{Cos ACD})$$
$<ACO + <ACB = 180°$
$\text{Cos}<ACB = -<\text{Cos ACO}$
Example $\text{Cos } 30° = 0.866$
$\text{Cos } 150° = -0.866$

Comment: This is a nice easy formula to use except angle ACD is not one of the angles of the triangle. It would be easier if in place of angle ACD we could substitute angle A of the original triangle.

Notice the coordinates of point A are (x,h) and the R value is b.

The definition of cos is x/r. In the above case the Cos ACB = -x/R, which in the original triangle can be written as CosC = -x/b. Since -x/b is the opposite of x/b, this means CosC is the opposite of cos ACD.

$$\text{Cos C} = - \text{ Cos ACD}$$
$$\text{Substituting:}$$
$$c^2 = a^2 + b^2 - 2ab(CosC)$$

Hence, we have the Cos Theorem!
Cosine Theorem 20: If the sides of a triangle are a, b, and c then:

$$c^2 = a^2 + b^2 - 2ab(CosC)$$
$$or$$
$$a^2 = b^2 + c^2 - 2bc(Cos\ A)$$
$$or$$
$$b^2 = a^2 + c^2 - 2ac(CosB)$$

The formula contains only the quantities or variables involved in the problem. This formula, can be solved for the cos of the angle or any one missing variable. Plus the fact, it is easy to remember.

Now back to the problem. The sides were 5, 7, 9 and we need to solve for an angle.

Draw the triangle and label the sides and angles.

$$c^2 = a^2 + b^2 - 2ab(CosC)$$
$$c = 9 \quad a = 7 \quad b = 5$$

Substituting:
$$81 - 25 - 49 = -2(5)(7)CosC$$
$$7 = -70CosC$$
$$-7/70 = CosC$$
$$-.1 = CosC$$

Using your calculator angle C is 95 degrees.

For angle B write the formula in this form:

$$b^2 = a^2 + c^2 - 2ac(CosB)$$

$$49 = 25 + 81 - 90CosB$$
$$49 - 25 - 81 = - 90CosB$$
$$-57 = 90cosB$$
$$-57/-90 = CosB$$
$$.633333 = CosB$$

Angle B is 51 degrees.

To find angle A subtract the sum of angles C and B from 180. mA = 34 degrees.

You can see why so many persons in professions with a math background are paid more money! The mathematics they need is more complicated. You may never use these two theorems or you may use them quite often in you trade or profession. These two theorems should be in your notes and when you need to work a problem involving Trig check your notes and follow them carefully.

Activity
Be Selective

For the following problems, use your notes, the text, and your calculator. Suggested steps to help you solve the problem are:

- Read the problem carefully!
- Draw the figure and label it.

Write the Sin and/or Cos theorem and substitute values.

Solve if possible (Some problems may not have solution.) Answer the following question. Does the answer seem logical?

Be selective and work the ones that are of interest to you!

---

1. Some student asked the following question: What is the measure of the smaller angle the diagonals of a football field form? The field is a rectangle 100 yards by 50 yards.

   a. Draw the picture and decide which angle you want to solve for.
   b. What is the length of each diagonal?
   c. Do the diagonals bisect each other?
   d. What are the lengths of segments AE and BE where E is the midpoint of the diagonals, A and B are the end points of a diagonal?
   e. Use Cos Theorem to solve for the
      a. smaller angle formed by the
      b. diagonals. (Answer to nearest
      c. degree.)
   f. What is the measure of the larger angle formed by the diagonals?

Answers: b. 111.8 yds   c. yes
          d. 55.9 yds   e. 53° f. 127°

Trick? This is not a triangle!
Why? 30 yds across.
Answer: 4 degrees on each side of the centerline for a total window of 8 degrees.

Ramp 7.9
Proofs

These theorems are not usually taught in a high school geometry course.

The author assumes you are interested in how this formula (V = 1/3 Base(area) times was derived. So, the following will walk you thru a proof.

Most geometry courses, except in college, do not justify a few key volume theorems. (You will enjoy this!)

The Justification:

Draw 3D picture of a cone
(Use the pyramid proof.)

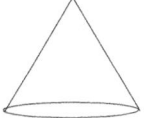

To prove this new theorem, we will use
the theorem for the volume of a pyramid
which is volume = 1/3 area of base
times the height. The limit of the area
of the inscribed triangles in the base
which is the area of the circle.
(THINK ABOUT IT.)

Two special and useful theorems which are not usually taught in a geometry course.

Basketball Pole Problem

(Also, the Telephone pole Theorem. All should understand this justification!)

A father and son were planning to erect a pole in the backyard to place a basketball rim and backboard on. They wanted the pole to be perpendicular to the plane determined by the cement drive way.

Theorem states that if the pole is perpendicular to 2 intersecting lines in the plane, then it is perpendicular to all lines in the plane containing the point. The son explained the method to determine perpendicularity to his father. What a geometric theorem is, the son explained to his father.

a. Draw a diagram to illustrate the theorem or make a model.
The proof is labeled Theorem 22.

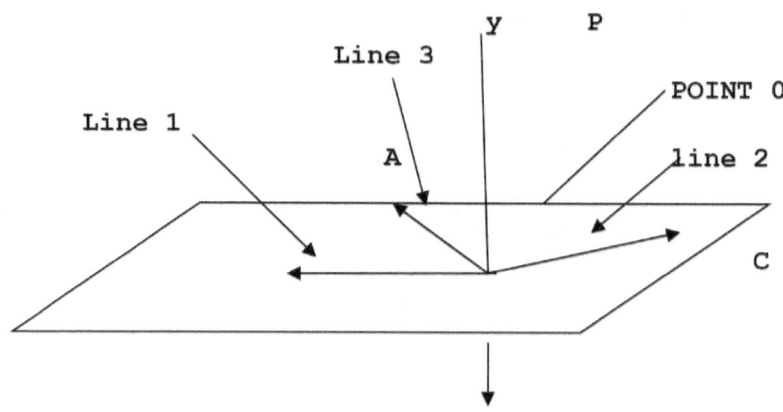

b. Line 3 and axis y form a plane and pick a **point(p)** on line 3 and construct a circle with a radius from the **point p** to **O** (add the circle to the drawing) What can you say PO is in regard to the circle? (They form a tangent.) Therefore, the line 3 is perpendicular to the plane. Which is what was to be justified. **QED**

All should know this theorem.
The other interesting theorem is the
Jordan Curve theorem.

Basically it points out that if a line divides a sheet of paper into two parts (A,B) then if you cross the line once, then you are on the other side (say B) and if you cross the line again, then you are back on the side A.

Conclusion: If you are on side A and cross an odd number of times the you are on the B side, or if you cross an even number of times, then you are on side A. **Call this the Jordan Theorem 22.**

Many State Fairs have such a building where people enter and try to find their way out! The easy way to find a way out is to keep your right or left hand on the wall and it will guide you out. Here is an example for you to try!

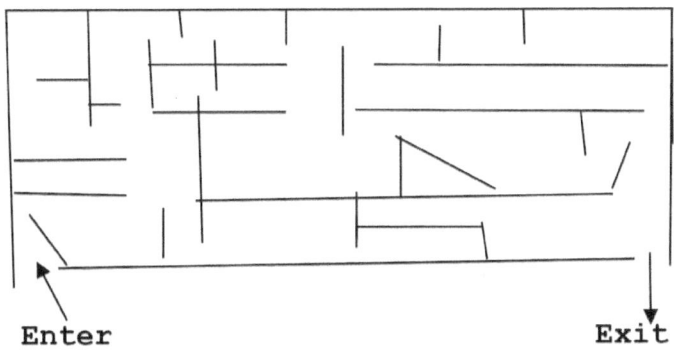

Enter   **Exit**

Hint: Keep either the right or left hand on the wall. This will lead you out!

Applications:
International Angle Measurement System

Comment: It is suggested that this be an Activity and explained by one of your group.

This activity may seem odd, but it leads into understanding of another system for angle measurement. A system that is used in the scientific world and in most of the countries of the world. The name of this international unit is **Radian.**

a. What is the circumference of the circle in "a"?

Answer: C = 2πr

b. How many times will the length of the radius divide into the circumference?

Answer: 2π times.

c. Draw on your circle in "a" with a central angle that intersects an arc on the circle equal to the length of one radius.

d. Is the measurement of the central angle in "d" equal to, larger than, or less than 60 degrees? Justify your answer.

Answer: Less than.

e. Calculate the measure of the angle in "d" in degrees and minutes.

Answer: 57 degrees 18 minutes
(Rounded)

f. The name for the measure of the angle in "d" is one radian. How many radians are in 360 degrees? How many degrees are in one radian?

Answer: 57 degrees 18 minutes
(Rounded)

How many radians are in one degree?

Answer: .017

Notes and Comments

Ramp 7.10
Trigonometric Identities

"Mathematics through the power of computers pervade almost every aspect of our lives..."

David L. Goines

After completing this section, you will:

1. **Know basic Trigonometric identities,**
2. **Justify a few basic identities.**

This Bridge may be exciting for you, if learning new ideas are exciting. To continue with new ideas, the word, Identity, needs to be defined. Just interpreting the root word, identity or identical, gives you the idea of sameness. It may remind you of congruent figures! This idea is correct, but if we apply this meaning to Algebraic or Trigonometric equations, the interpretation is a little different.

Activity

1. What is the difference between these two equations?

$$3x + 15 = -6$$
$$2(x + 3) = 3(x - 1) + 9 - x$$

Hint: Try to solve each equation.

The first equation has a unique value for x, or a unique answer, x = -7. The second equation has many answers, in fact any number will check in the second equation. Try substituting a few values for x in the second equation.

The second equation is an identity!

Another geometric identity, that you have investigated, is the Pythagorean Theorem. It is always true for a right triangle.

2.The following values and write a conclusion. (Use your calculator.) Substitute these values for A in the above highlighted expression. (All are in degrees.)

     a. A = 30 b. A = 45 c. A = 60
     d. A = 90 e. A = 120 f. A = 135
     g. A = 150 h. A = 180 i. A = 210

Definition 33: An equation is an identity if the equation is true regardless of the value that is substituted for x or the variables.

Identities
The Basic Trigonometric Ones

Since it has already been stated that the Pythagorean Theorem is an identity, we just as well show the trig form of this important identity. Again, the only information we can use is the information we already know. (This sounds a bit odd, since naturally, we can't use information we don't have.) The information needed is:

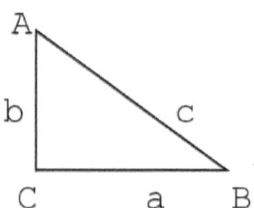

     Sin A = a/c   Cos A = b/c
     Tan A = a/b

We also know the important Pythagorean Theorem, a^2 + b^2 = c^2, is valid for any right triangle with sides a, b, and hypotenuse c.

Do you think that
  1. Tan A = Sin A/Cos A
     Tan A = a/b,
     Sin A/Cos A = (a/c)/(b/c) = a/b

  Which was to be shown!

  2. Sin² A + Cos² B = 1?
     (A/C)² + (B/C)² = 1
     A² + B²= C²

  Q.E.D.

This can be written in a shorter way as follows. (Notice the placement of the exponents.):

$$Sin^2 A + Cos^2 A = 1$$

This is the Trig form of the Pythagorean Theorem – a very important identity!

Theorem 24:
Given any angle then
Sin² A + Cos² A = 1.
(Enter this in your summary.)

Example:
     Given A = 60 degrees
        $Sin^2 60 + Cos^2 60$= 1.
     Using a calculator
        $.866603^2 + .5^2$ should equal 1?
        .75 + .25 = 1

     If the exact values are used we have:

$$\sqrt{3}/2)^2 + .5^2 \text{ should equal } 1?$$
$$3/4 + .25 = 1$$
$$1 = 1$$

There is one more identity that is important for you at this time. It is an easy one to derive and to remember. Remembrance is related to use and the more you use an idea or a word the longer you will remember it. (Like the combination for your locker.)

This is the second important identity, and we will call it a theorem.

Theorem 25:

If A is an angle then
Tan A = Sin A/Cos A
for values of A except 90,
and odd multiples of 90.

Comment: Cos A cannot be zero or A cannot be 90, or an odd multiple of 90 degrees.

Examples: These are the only two trig identities that are important at this stage of your studies. I almost used the word game, because now you will be asked to work an activity, which resembles a game. The only two players are these two identities and you, plus your algebra skills.

Activity

1. Is Tan A times Cos A equals Sin A? (This problem is asking you to determine if the product of Tan A and Cos A is equal to Sin A.) We could try a few angles to get an idea if it is true, but this would only give us verification for a few cases. The justification for an identity should show it works

for all cases. We will use the trig identity: Tan A = Sin A/Cos A and substitute this for Tan A.

$$(Tan\ A) \times (Cos\ A) = Sin\ A$$
Is this an identity?

Given: **Tan A = Sin A/Cos A**
Substituting <u>Sin A/Cos A</u> for Tan A
in the original question
(Sin A/Cos A)Cos A = Sin A? The Cos
A divides with Cos A leaving:
Sin A = Sin A

Q.E.D. The original was converted to Sin A = Sin A and therefore, it is an identity!

Comment: Check in a dictionary for the meaning of Q.E.D., if you have forgotten. Problem 1, you might be thinking, this was too simple! So, try number 2.

2. Is this an identity?

$$(Sin\ A) \times (1/Tan\ A) = Cos\ A$$
Substitute Sin A/Cos A for Tan A and simplify.

Some students really like proving identities and some don't. It appears to be a like or dislike type of problem. I hope you are in the "like" group and will even ask for more exercises once in a while. This Bridge contains two theorems, which are listed below.

Definition 33: An equation is an identity, if the equation is valid regardless of the value that is substituted for x or the variable.

Activity
Try a few
(Optional)

Prove the following identities! (These all are Valid identities.) Some students really enjoy these identities.

Hint: Convert some of the cases to Sin and Cos and then simplify.

1. Cos A = (1/Tan A) Sin A
2. (1/Cos A) - (Tan A)(Sin A) = Cos A
3. Sin A)(1/Cos A)(1/Tan A) = 1
4. Cos A x Tan A x(1/Sin A) = 1
5. $1 + 1/Tan^2 A = 1/Sin^2 A$
6. $Tan^2 A - 1 = 1/Cos^2 A$
7. (1/Cos A)(1/Sin A) = (1/Tan A)+Tan A
8. (Tan A x Sin A) + Cos A = 1/Cos A
9. $Cos^2 Ax (1 - Tan^2 A) = 1$
10. $Cos^2 A - Sin^2 A = 2 x (Cos^2 A - 1)$
11. Tan A x Cos A = $\sqrt{1-Cos^2 A}$

Now you know these few identities, since you may have justified them, and they should work for all values of A (any angle). Try to check Tan A x Cos A = Sin A for A equal to 90 degrees using your calculator.

Comment: Calculators may give "error" for the left side and 1 for the right side. Looking more closely, the lefthand side reads E times 0 equals the right side, which is 1. This is why the restriction is imposed that discontinuous functions or certain values such as 90 degrees (or an odd multiple of 90), since the Tan function is undefined at those values or division by 0 is not possible.

Investigations-Optional
All should try these but be selective

Two sides of a triangle are 1, $\sqrt{3}$.
The angles of the triangle are ?, ?, ?.
What are the "?" values?

   a. Can you make a right triangle with the lengths of two sides?
   a. Draw the triangle and label it.
   a. Calculate the measured of the angles.
   b. Answer the question in the problem.

Test: Beberman's friend
Interesting and Enjoyable

It was mentioned that Mathematics somewhat resembles a game and in order to play a game you must know the rules. The following example was created by Prof. Beberman from the University of Illinois as an example of how students may misinterpret problems. Your friends may enjoy this?

"Stan Brown had a pen pal, Al Moore, who lived in Alaska. Stan and Al corresponded quite frequently. Stan liked to receive letters from Al because he wrote about interesting things like hunting and fishing and prospecting for gold. Al enjoyed hearing about the things Stan did, especially about school. One day, Al wrote to ask if Stan would mind teaching him some arithmetic. Stan agreed but decided he needed to know how much Al already knew. So, in his next letter to Al he included a simple test, and asked Al to write in the answers and to return the test to him."

Below is the test Stan sent. Work each problem (You may want to ask your friends to work the test also.) and record their answers.

Test
What are your Answers?

1. Take 2 away from 21.
2. What is half of 8?
3. Add 5 to 7.
4. Which is larger, .00065 or .25?
5. Does 2 x 4(1/2) = 9?
6. How many times does 3 go into 8?
7. How many times does 9 go into 99?
8. Which is larger, 3 or 23?
9. Write a number smaller than 4?
10.Write a number larger than 4?

Here are Al's answers!
(Stan was flabbergasted, are you?)

1. 1	2. o
3. 57	4. .00065

5. No (Al did not read x as times.)

6. Twice	7. Twice
8. 23	9. 5

10.3

Explain Al's interpretation for each question.

Create some other examples. Here are some hints.

1. Half of 8 is 3 (the right half)
2. Half of twelve (XII) is seven (VII). The upper half!

Create some of your own!

Ramp 7.11
Trigonometry Applications Review

"Mathematics is not a spectator sport!"

unknown

This Bridge began with an anonymous quote which appeared contradictory, and hopefully now makes more sense to you. There, possibly, hasn't been a of difficult new material in this Bridge, but it is different and will probably require you to refer to your notes quite often. There is nothing wrong with this! It is a characteristic of a good student. Trigonometry is used in the fields of science, electronics, music, construction, navigation (land, sea, air and space), and in recent years the medical field. Many of the real applications require more math skills then you have at this time, so the examples presented here are elementary. Being a thinking person, you have probably questioned where the names Sine, Cosine, and Tangent came from. The word SINE evolved from the Hindu word related to curve and the term COSINE came from the relationship to the compliment of the angle, hence Sin A and Cos A were abbreviated to Sin A and Cos A. This will be clearer as you complete this section. Exercises 14 and 15 in Activity 6,6,2, will clarify the reason for the definition of Tangent A.

Activity
Use your calculator
A good problem to work by a group

This curve has many applications!

1. On a sheet of **graph paper** draw the x and y axes as in the figure below. From the origin point, (0,0), the cm divisionsn10 to 90 on the positive x-axis.

The y-axis will have 1 unit equal to 10cm. The graph of Y = Sin X, for X from 1 to 180 degrees. (every 15 degrees)

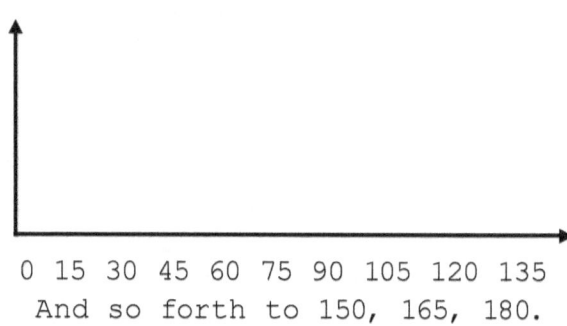

0 15 30 45 60 75 90 105 120 135
And so forth to 150, 165, 180.

2. Complete this table and plot the points on the graph. Then draw curve though the points.

X(in degrees) y = Sin X

X	Sin X	X	Sin X
0	0	90	_____
15	.17	105	_____
30	.5	120	_____
45	_____	135	_____
60	_____	150	_____
75	_____	180	_____
90	_____		
105	_____		

y = Sin x for x from 0 through 180 degrees. Notice, it is a curve related to the meaning of the word SINE, which means a bow. Keep this graph, it will be used in exercise 8 in the home activity. Now using different colors plot the values for the Cos and Sin curves, but continue the graph for 0 to 180 degrees. **These are the curves for alternating current electricity.**

1. Sin = ?
2. Cos = ?
3. Label the figure and fill in the given information.
4. Identify what you need to solve for.
5. Try the Sin or Cos theorems since these will solve any triangle that can be solved.
6. Answer the question.
7. Round the answer to a reasonable value.

(Ask an electrical engineer to explain W = VA(cosA).

Work for your group!

Review Problems
Be Selective

1. The Tax Assessor needs to know the area of the lot in order to determine the taxes. The sides are 1000 by 600 by 500 feet.

a. The city engineer wants to know the largest angle. What is the measure of the largest angle to the nearest degree?

Hint: Cos Theorem

b. What is length of the altitude to side that measures 1000 feet?

c. What is the area of the triangle?

d. How many acres are in the triangle? (1 acre=43560 sq. ft.)

e. If the tax on unimproved city lots is $300 per acre or fraction thereof, then what is the tax on this lot?

Answers: a.130.5 b.227.7 c.113836.8 sq. ft. d.2.6133 e.$783.99

Comment: This is a good problem to illustrate the different dollar and cents answers the owner could be charged depending on how the number of acres are rounded off. What is the tax per acre? (Or square yards in your area

Answers: a. 4063 sq. ft. (4062.3 rounds to 4063. Why?) b. $677.17

2. An escalator should not be inclined more than 25 degrees. If an architect plans to lift people 20 feet to the next floor and use the angle of 25°, *then:*

a. What will be the length of the escalator?
b. What horizontal length of floor will be needed? (Answer to the nearest foot.)

Hint: Right triangle, use Sin and Tan functions.

AE=ED,
BA=BD,
Angle BEC = 90 degrees,
BE = 30 yards,
AD = 24 ft. and
CE = 10 ft.

Draw the figure.
Answers: m∠CBE = 6.3 degrees
m∠ABD = 15 degrees

Take a break!!!

Bridge 8 Statistics

Ramp 8.1
The Use and Misuse of Statistics
(A VERY IMPORTANT BRIDGE)

One night per week during and/or after dinner time for family discussions. Topics: retirement, when, why, and where.

"A mathematician, like everyone else, lives in the real world. But the objects with which he works do not. They live in that other place the mathematical world. Something else lives there also. It is called TRUTH."

Jerry P. King
THE ART OF MATHEMATICS

A few questions: What are the definitions of MODE, Median, and MEAN? Why and how are the three different averages used in society? You probably knew these terms at one time. Can you interpret these graphs, better still, do you recall their weaknesses? Plus interpret statistical graphs?

More Questions:
- Should teachers grade on the curve?
- Do news reports provide the proper or sufficient information for the implied decisions?
- What is the normal curve the bell shaped curve, and how is it used?

The answers to these questions and many more are needed for effective citizenship. They will be explained and practiced in this Bridge.

This could be the information for the family to discuss this, especially if there are children in school or college.

Ramp 8.2
The Meanings of the Three Averages (Mean, Mode and Median)

After completing this Bridge, you will understand:

1. Definitions of mode, median and mean.
2. Calculate or determine the mode, median, and mean for a set of data.

The calculating of averages is fairly simple, but interpreting the results accurately can be confusing and is often misleading. You may come across numbers representing averages in the news every day and as an informed citizen you should be able to understand the correct meaning of these. You have probably heard the statement: **"Figures don't lie, but liars' figure."** It is important for you to understand the use and misuse of these averages and their implications. This Bridge will help you interpret statements involving statistics and other information that is needed to see the true picture.

The reporting of scores or sets of data to convey and understand the results or the graph by using averages is often used.

Some examples of these are:

a. The meaning of math scores for a certain grade or class in your community.
b. The average weight of the Green Bay Packers defensive line.

c. The average score per game for your school's basketball team.

d. The average stopping distance for a certain make of car.

e. The average speed for a trip.

f. The average for your school's ACT or SAT scores.

g. The meaning of political surveys.

Activity

1. Let's say you are driving from Missoula, MT to Billings, MT a distance of approximately 350 miles and it is all interstate highway. On this section of freeway there are no stoplights so once on the freeway you can set the cruise control and not stop until you are at your destination, unless you wish a rest stop.

Given the following graph, explain why the average speed was approximately 50 mph, and only slightly higher if the 20-minute rest stop is not included in the time. You may need to use the following definition.

Definition 39: The average speed is the

distance divided by the by the time it takes to travel the distance.

Average speed = D/T.

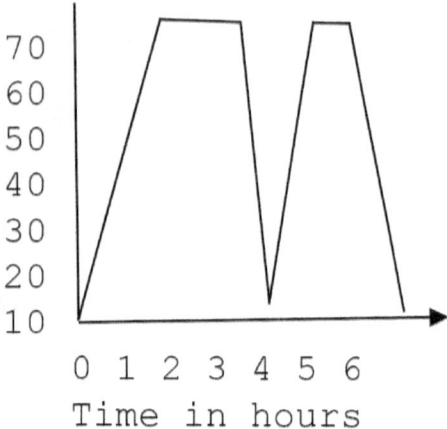

Time in hours

From the graph, estimate the time the car was actually traveling 55 miles per hour? Where is the rest stop indicated? What could the driver do so that the average speed is nearer the speed the cruise control is set at?

Estimate: what is the average speed for the trip? What would the average speed have been without the rest stop(estimate)?

Answers:
- Estimated answers are 60 mph for approximately 3.5 hours.
- Rest stop is between hours 2 and 3.
- Save time by taking a shorter rest stop or no rest stop.
- 55 mph is the estimated average with rest stop.
- The average speed without rest stop is approximately (? mph)?

2. Let's look at the arithmetic average of the following Set of scores on an eighth-grade math team exam.

Scores: 100, 99, 80, 76, 70

The superintendent proudly reported to the newspaper that the average score for the eighth grade 2012 math team is 85. Does this one number give a true picture of the team's scores? Notice, no student even scored an 85.

Definition 40: The MEAN or arithmetic average for a set of number is the sum of the numbers divided by n, the number of items in the set.

Formula: Mean = (sum of n scores)/ n

Calculate the mean for the above eighth grade math team set of scores and check if 85 is correct.

The 2021 and 2020 math team scores are listed below. The superintendent reported the team had the same Average in each year:

2021 scores: 100, 99, 80, 76, 70
2020 scores: 100, 95, 91, 80, 59

Calculate the mean score for each year.

In both cases only the average (85) was reported to the parents. Was he correct?

Do you think the two teams are the same with regard to their accomplishment? Which team would you say is just one number reported for each of the two teams?

Comment: Discuss the above situation with the family. Now to further confuse the interpretation of data, there are three types of averages, mean, mode and median. Each is use by the business world, in education, the news media, the financial world and in the medical world. The mean is defined above and the mode and median will now be defined.

Definition 41: The MODE for a set of data is the most popular or most frequently occurring element or score in the set.

Definition 42: The MEDIAN for a set of data is the middle element or score when the elements or scores are arranged in order of size or quantity.

These three, mean, mode and median, are many times called the average, but the one that is being used is not always reported. Each of these averages, will be further clarified and compared, by applications.

3. Here is a set of income figures representing the annual payroll in a small business, including the owner's salary.
$18,000, $18,000, $18,000, $22,000, $25,000, $25,000, $80,000.

What do you think is the salary of the owner? Why?

Answer: $80,000

Using definitions calculate the three averages.
The Mean is _____. The Median is _____.
The Mode is _____.

Answers: Mean = $29428.57
Median = $22000
Mode = $18000

Each of these were reported as the average! Which average do you think the owner claimed for the average pay? Why?

Answer: Mean or $29428.57

Which average do you think the employees claimed for the average pay? *Why?*

Answer: Possibly the mode, $18,000.

4. What is the mode and median for each of the 2020 and 2021 math team scores?

2020 scores: 100, 99, 80, 76, 70
2021 scores: 100, 95, 91, 80, 59

Answer: No mode in either set. Median is 91 for the year 2020 team and 80 for year 2021.

Activity
Be selective

In the following calculate and identify the three averages. Use your calculator and/or computer. Each person takes a few and be selective.

A Geometry teacher was giving four geometry classes the following:
    Class 9 is at 9 AM.
    Class 10 is at 10 AM.
    Class 1 is at 1 PM.
    Class 2 is at 2 PM.

The four classes are to use the same textbooks and have the same final exam questions, but the individual problems will be in different orders for each class.

Questions:
- Does the time of day make any difference as to the class averages?
- Do you have any other possible conclusions?

An A on any exam A question is worth 4 points.
A B on any exam B question is worth 3 points.
A C on any C problem is worth two points
A D on any D problem is worth one point.
An F on any F problem is worth zero points.

1. First, arrange the questions in a ascending working order. In other a easy working order. (These 4 classes gave the same test for the same course but, a different order.)

Class	Grades	A's	B's	C's	D's	F's	Total Points
9 AM	Tests	2	8	7	5	0	
	Points	8	24	14	5	0	51
10 AM		10	9	3	1	6	
	Points	40	27	6	1	0	78
1 PM		1	11	2	4	2	
	Points	4	33	4	4	0	45
4 PM		4	13	8	3	1	
	Points	16	39	16	3	0	74

Question: The teacher recorded 2As and 2Bs for the final grades. What do you think? What do you think or would record? The teacher knows the students and their past their work the past scores, grades and work for classes 10 and 4.

2. Many times scores are grouped for easier calculating of the averages and for graphing. (Graphing of a few scores will be studied in the next activity.)

3. Montana is a very large state, approximately 700 miles wide and 400 miles deep, yet the evening national weather news will assign one number to indicate the temperature for the whole state. The following numbers indicate the variation of temperatures comparing five of the larger cities in Montana in August. What do you think the news media reported for the State of Montana with these temperatures for the largest cities:

101, 95, 67, 45, 36?

a. What is the mean temperature for these readings?
b. What is the median temperature for these readings?
c. What is the mode temperature?
d. What would you report for the state of Montana?

Answer: There is no correct figure for the State's temperature.
   a. Mean is 68.8
   b. Median is 67.
   c. There is no mode.
   d. Just report the high and low and where these are.

4. Baseball players keep track of their batting averages. A player has an average of 290 after 75 times at bat. What will he have to average in the next 40 times at bat to have an average of 300?

Answer: 318.75

5. A baseball player's batting average is 280 in 30 times at bat. How many hits will he need in the next 80 times at bat to average 300 for the 110 times at

bat? Estimate the answer before you calculate the correct answer.

> *Answer:* Approximately 30 (30.2, but it is impossible to hit a .2 hit).

6. Many colleges and some high schools use A = 4, B = 3, C = 2, D = 1, and F = 0 to calculate grade point averages (GPA). If a student has a 2.0 average after 8 courses, then what must he average in the next 24 courses to average 3.5. Guess first!

> *Answer:* 4 or all A's.

7. Another student at the same school as in #6 has a 2.0 average after 16 courses, then what must she average in the next 16 courses to average 3.5. Guess first!

> *Answer:* **It can't be done!!**

8. From results of problems 6 and 7, what is your conclusion as to when the high grades should be made?

Comment: Discuss your conclusions with your friends and arrive at a few conclusions.

9. Toss a pair of dice 36 times and keep track of the number of times the various sums of the two faces occur.

What are the theoretical sums and the probability? Complete the table:

Sum	2	3	4	5	6	7	8	9	10	11	12
Probability	1/36										

a. What is the theoretical probability for throwing each of the sums above?

b. What should the theoretical probabilities add to?

c. From your 3 tosses, what is the empirical probability for throwing the following sums? 4? 7? 12? 1?

d. What is the probability of throwing a 5 **and** then a 7? (In this case, the "and" means multiply the two P(x).)

e. What is the probability of throwing a 5 **or** a 7?

In this case, the **"or"** means to add the values of the 2 dice (Px) and the **"and"** means to multiply the value of the 2 dice. This a legal definition for the use of the words.

Ramp 8.3
Investigations of the Geometric Mean
Be selective

1. Draw a three-inch line segment. Label the endpoints A and B. Label points M and M1 one inch from A and B on segment AB. Using AB as a Diameter draw 2 circles with radius AM and at centers BM1 the circle. Label the center of MM1 as 0. At M construct a perpendicular that intersects the circle at C and connect points A and C, and B and C.

a. What kind of a triangle is ABC? Why?

b. Prove: $CM = \sqrt{(AM)(BM)}$

Answer: b. Map the similar triangle and write the proportions, then selected the proper ratios and justify the following. Therefore: CM =The sq. rt. of (AM)(BM).

Application
The Optic Nerve problem

2. Take a sheet of paper and near the center make two solid circles or dots about a ¼ in. in diameter and 2 inches apart. Hold the paper a few inches from your face and close or cover your left eye. Now look at the left dot or circle with your right eye and slowly move the paper straight away from your nose until an arm length or that the right dot disappears. It may take several tries to observe this result. Stop moving the paper when the right dot disappears

   a. How far is the paper from your eye?
   b. Repeat the process with the other eye covered.
   c. Is the distance the same?
   d. Explain the disappearance of the dot.

   (You may need to view an eye diagram, an article in a encyclopedia, or a doctor.)

Thinking
Measures of Central Tendency

3. a. Why are the three averages call the "Measures of Central Tendency?"
b. Which average best describes a characteristic (such as weight or test average) for large sets of data. Why?
c. Why is it incomplete for schools to report SAT or ACT average as a score to support excellence?
d. Research: What percent of the seniors in your school took the SAT last year? What was the average score?
e. Research: What percent of the seniors in your school took the ACT last year? What was the average for each of the five reported scores?

Answers:
>   a. The scores indicate how centralized they are.
>   b. Probably the mean, if the scores are not bias or if the set is well defined and large enough.
>   c. Only selected students take the tests, and the number is usually is not reported. The schools, in most cases, report only the composite scores and not the scores for the various sections of the tests. Nor are the tests curriculum results used to improve the program.

Complete your notes.

The one per week discussion could report the history behind the school system, when, how, where and future.

Ramp 8.4
Graphs and Interpretations

"You cannot fake in mathematics, no one can be fooled. You can either prove (or solve) or you cannot."

Jerry P. King
THE ART OF MATHEMATICS

After dinner project information for the family – make a graph that compares State or Federal Income Taxes you paid over the last years.

Types of Statistical Graphs
(The Picture Story)

After completing this review section, you will better understand:

1. Methods for constructing the various types of graphs.
2. Interpreting and reading graphs.
3. Conclusions from a graph and understand the weaknesses of the graph.

There is a saying: "A picture is worth a thousand words." What this section will do is help you review different ways to represent a set of data. In Ramp 1 of this Bridge, you found the three averages known as measures of central tendency. You also discovered that three averages do not tell the whole story. In fact, they can be misleading and possibly lead to false conclusions. To prevent false conclusions a better picture method is required. There are several types of graphs to represent data, but in each the axes must be properly labeled, a correct title, and the graphs must be easily interpreted. Many times, color is used to provide a "colorful" aid to interpretation and attract attention. Some of the types of graphs are reviewed easy interpretation. The same data is used for the types of graphs. Some of the names given to the types are: *Bar, Line, Circle, even three dimensional* and each of these are usually uniquely adapted to the material or subject of the graph and hopefully attracts attention.

Scores

Notice each graph must have sufficient information on it to enable it to be interpreted, or "tell the story!" Which type do you feel is the easiest to interpret?

Comment: The circle type seems to be the hardest for most people to construct. (If you can find a few graphs in the paper or magazines cut them out use them.)

Activity

1. The following are grades on a test in Physics rounded to the nearest multiple of ten. Mr. W's Physics Class:

Grade	100	90	80	70	60	50
Frequency	2	12	6	3	4	2

    a. How many students are in the class?
    b. What is the mode? mean? median?
    c. Divide a sheet of paper into four equal sections and construct four different types of graphs. (Suggestion: one type per quadrant.)
    d. Indicate the three averages on each graph.

Answers: a. 29 b. Mode is 90 Mean is 81. Median is 80.

2. The following shows the salary per employee per year for a small business.

Salary (in Thousands)	Number of employees
$16	
$21	11
$24	8
$29	6
$38	4
$41	3

$53	2
$162	1

a. How many employees are there?
b. What is the mode? mean? median?
c. Divide a sheet of paper into four equal sections and construct four different types of graph (one type per quadrant).
d. Indicate the three averages on each graph.
e. Which average most accurately describes the average salary? Why?
f. What is the mean if the $162,000 salary is not considered?
g. Which is the salary of the CEO?

Answers: a.42, b. Mean = $30,000, Mode = $21,000 Median = $24,000, c.$27,000 rounded.

Application
Center of gravity

1. a. Locate the center of gravity for a ruler by the balancing method.
   b. Locate the center of gravity point on a pencil or pen by balancing.
   c. Locate the center of gravity point on a baseball bat by balancing.

Comment: The center of gravity is the point where the most energy is transferred to the ball, or the point where the batter hopes the bat will make contact with the ball. It is also called the sweet spot. (Ask your coach.)

Where is the sweet spot on a golf club?
Write your summary.

Ramp 8.5
The Normal Curve

"Statistics makes possible new perceptions and realities by making visible large-scale patterns."

Neil Postman

TECHNOPOLY

(Suggestion: Read Technopoly, yes just read it carefully.)

This Bridge will help you:
1. Better understand the meaning of "normal curve."
2. Better understand why the curve is called the "normal curve."

As a student, you have no doubt heard of the term "normal curve" and/or have asked the teacher the question, "Are you grading on the curve?" Did you really understand the meaning of the term and the consequences of the question? If the answer was no, then this section will help you understand the meaning, the derivation and the use of the term normal curve. It is used, very often, in business, education and the sciences.

A bit of history: the term normal curve was really a by-product of World War I. The curve was first introduced by Abraham De Moivre (1667–1754) and Karl Gauss (1777–1855). Sometimes it is called the Gaussian Curve. The U.S. Army began to draft men and naturally had to provide the recruits with wearing apparel such as shoes. One procedure would be to measure the size of each soldier's feet and then order the shoes. Naturally, this would take time and create an awkward situation as the soldiers waited for shoes. Another option would be to order large quantities of shoes in all possible sizes. This would probably leave a number of pairs not used like the very small sizes or the very large.

Research: Karl Gauss and Abraham De Moivre

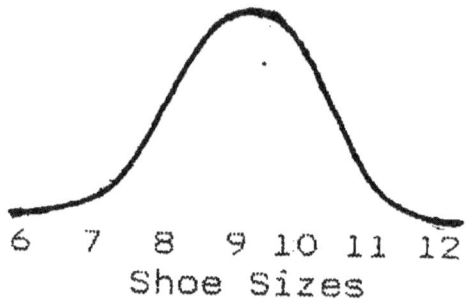

6   7   8   9 10 11 12
Shoe Sizes

From the curve, an estimate can be derived as to the number of pairs to order of each size. You can see more 9s would be ordered than any other size. This is what the army did in order to have the sizes available for the recruits. This procedure was also applied to other equipment needs where size is a factor.

The NORMAL CURVE is also known as the BELL-SHAPED curve. The name **"bell shaped"** is due to the shape of the curve. It is also identified as the "natural curve" due to the fact that many patterns in nature resemble the curve.

This curve has several important properties and many different applications (life, growth, medical, ability, weight, income, memory).

Properties of the Normal or Bell are shaped curve quite equal: The mean, mode, and median are all similar and are on the line or near the line of symmetry. Plus or minus one STANDARD DEVIATION determines the amount of spread on each side of the mean or line of symmetry plus it will include approximately **68%** of the data.

1. Two STANDARD DEVIATIONS on each side of the mean (line of symmetry) will include approximately **95%** of the data.
2. Three STANDARD DEVIATIONS on each side of the mean (line of symmetry)will include approximately **99.8%** of the data.
3. From the data or the curve the RANGE (The measure from the lowest to the highest) can also be determined.

The above is summarized on the graph below using the shoe sizes.

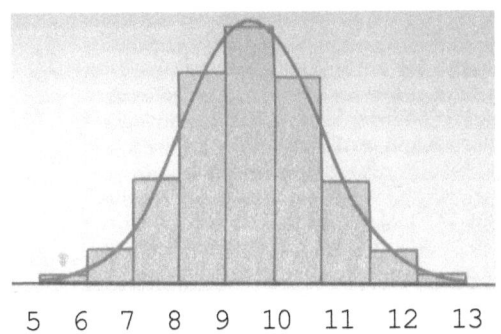

5 6 7 8 9 10 11 12 13

Your question no doubt is: How is the value for STANDARD DEVIATION calculated? What we need are some definitions!

Definition 43: The RANGE is the difference between the highest or largest number and the lowest or smallest number in the set of data.

(In the shoe case above, the range is 5 to 13.)

Definition 44: The Standard Deviation of a set of data is the square root of the mean of the scores.

You better read definitions a few more times in order to really understand them. The following formula for this definition is:

$$SD = \sqrt{\frac{(s_1-m)^2 + (s_2-m)^2 + (s_3-m)^2 + (s_n-m)^2}{N}}$$

Where m is the mean score, s is a unique individual score, and N is the number of scores.

You can see this takes some effort and time with your calculator. Some calculators have built in program for calculating the standard deviation. (Check the instruction book for your calculator or computer.) You also, no doubt, wonder where the percentages (68-95-99.8) came from (See definition: Normal Curve.) These are derived by using CALCULUS, a college course in mathematics and the numbers refer to the area under the curve. This means you will have to wait until you take the Calculus for the explanation and just accept it for now. (Call it an assumption if you wish.)

The following example will help you understand the Standard Deviation concept, provide an appreciation for the effort to calculate this information, and how it is used in interpreting the "picture."

Activity
Work selected problems

1. The following data are the results of surveying 100 male students as to their shoe size.

Size(S)	Frequency	%	N
6	2	2	6
7	12	13	24
8	19	19	19
9	29	21	88
10	12	13	21
11	15	16	30
12	3	3	9

a. The first chore is to calculate the three averages (mean, mode, and median).

Answer: The mean is 9.1 rounded to 9. Mean = 9, mode = 9, median = 9 (rounded) and the data is assumed to be normal.

b. Now using the definition, calculate the Standard Deviation or SD. You should arrive at an SD value of 1.4 rounded. (You may not want to calculate the SD.)

c. Draw the curve.

This tells that 68% of the men should take sizes 8,9, or 10. About 14% will take size 7 and the same number for size 10, with about 2% taking 6 or 12. My guess is if you were the buyer, you would order more 9's than any other size. It could also be assumed from the 100 cases the following:

> 2% will take size 6
> 12% will take size 7
> 19% will take size 8
> 28% will take size 9
> 21% will take size 10
> 15% will take size 11
> 3% will take size 12.

245

Now draw the graph using the above and include the lines segments showing the following.

    a. the mean, mode, median
    b. + and - 1 SD
    c. + and - 2 SD
    d. + and - 3 SD
    e. Include a title and label the axes.

Your graph should be similar to the following normal curve.

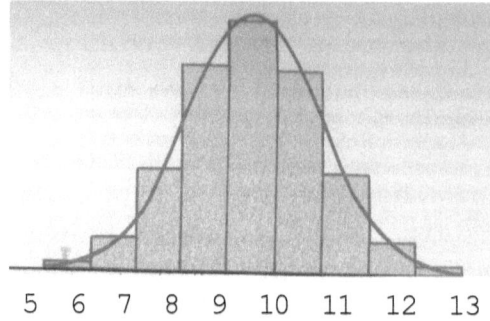

5 6 7 8 9 10 11 12 13

You can easily understand how such data and conclusions are used by the military!

Suggestion: ask a manager of a shoe store how they order shoes.

Some person once said, **"Mathematics is not a spectator sport."** This means in order to understand the concepts and learn the skills one needs to practice!

Record notes for Bridge 8.5.

A Family Topic

Plan for a family reunion in the future. (The author had one and over 100 family members attended!)

Ramp 8.6
The Use And Misuse Of Statistics

"Attributing teaching and learning failure to something called "math anxiety" serves no purpose except to provide a built-in excuse for inadequate performance on both sides."

Jerry P. King
THE ART OF MATHEMATICS

Review and Applications
Select the items of interest

It was H. G. Wells, who said:

"Statistical thinking will one day be basic and necessary efficient citizenship as the ability to read and write."

The continued education of people has not yet reached the goal stated by H. G. Wells, but this Bridge is a step in the right direction.

Information

Statistics is used to provide an explanation or picture of the past, partly based on the saying: a picture is worth a thousand words. This gives us insight to part of the problem. Different viewers will interpret what they see differently. The picture below points this out? Do you see a young woman or an old woman or both? This was created by W. Hill and appeared in Puck magazine in 1915.

Comment: Keep looking until you can see both the old and the young woman. Hint: The mouth of the old woman is the necklace on the young woman and the nose on the old woman is the jaw of the young woman. Look from the right and then from the left.

Statistical methods today are influencing the Decisions about the future in the medical, business, sports, agricultural, biology, economics, education, electronics, physics, psychology, sociology, chemistry, space programs, military programs, etc. H.G. Wells stated the need, but John Dewey states a concern: "Factual science may collect statistics, and make charts. But its predictions are, as has been well said, but past history reversed."

In this Bridge, you have been exposed to the following, which will be reviewed in the activities in this section.

Ramp 8.7
Descriptive Statistics

1. Graphs
   a. line segments
   b. bar
   c. circle
   d. icon
   e. normal curve
2. Measures of central tendency
   a. mean or arithmetic average
   b. mode
   c. median
3. Measure of spread
   a. range
   b. standard deviation
4. Polls
   a. random sample
   b. possible errors
5. Potential concerns as interpretations leading to errors.
   a. Were the questions ambiguous?
   b. Were the data random?
   c. Was the sample large enough and random?
   d. Were the generalizations valid?

More REVIEW!
Be Selective

If the answers are given you may want to used them instead of solving for the answers. Interpretations are more important than solving the problem!

1. Prepare a brief statement with some assumed answers:
   a. When did H. G. Wells and John Dewey live?
   b. What were their noted for?

    c. What did they mean by the statements that are quoted above?

2. It is estimated that:

    70% of the population have brown hair

    15% have blond hair.

    10% have black hair

    5% have red hair.

    a. Graph and Interpret the above using a bar or icon type of graph.

    b. Interpret the above using a circle graph.

    c. Which type of graph do you feel is easier to interpret? Why?

    d. Show your graphs to a few of your friends and ask which one they like and is easier to understand. Record your findings.

No single numbers like the mean have less meaning and are influenced by the extreme scores.

Identify which average is used. How will the individual scores be used to help the students? If this test has been given before? Were the weaknesses and strengths the same?

3. The following is partly from the September 10, 1997 Missoulian, the local Paper in Missoula, Montana.

Headline: **Young Americans are confident and optimistic about their future.**

Another poll: A national telephone survey of 2,001 teenagers concluded:

Eight out of every ten believe they will be more successful than their parents, when they reach the (parent's) age.

The article then used 12 inches of type explaining what the youth hope to do in the future and what they don't want to do. The responses were as you would expect: The teenagers want good pay, continue their education, job security, be creative, and avoid routine work.

What are some of the questions the article should have answered to make the surveys more creditable and informative or you would asked? Definition of terms such as routine work.

4. a. Write the next row of numbers in the following pattern.

```
              1
           1     1
        1     2     1
     1     3     3     1
  1     4     6     4     1
1     5    10    10     5     1
1     ?     ?     ?     ?     ?     1
```

b. Expand the a few of following.

$(A + B)^0 = 1$
$(A + B)^1 = A + B$
$(A + B)^2 = ?$
$(A + B)^3 = ?$

c. Do you see a similarity between parts a and b?

Comment: This is called Pascal's triangle and is a clever way to determine the coefficients for binomial multiplication.

Does this apply to probabilities? Yes, it does.

Take the case of tossing coins. If there is one toss of one coin then, the possibilities are:
Heads or Tails T or H
    Probability is (1/2 1/2)

If there are two tosses of one coin, the possibilities are:

TT   or **TH**   or **HT**   or **HH**
1/4   **1/4**   **1/4**   **1/4**

If there are three tosses of one coin, the possibilities are:

1 out of 8 for one of these combinations
TTT, TTF, TFT, FTT, TFF, FTF, FFT, FFF

How do the probabilities figure in? For one toss the probability for T is ½ and for H is 1/2. For two tosses the probability is:

    TT or P(2H) is 1/4, P(TH or HT) is 1/2,
        And for two Tails is 1/4.

Ramp 8.8
Implications, An Important Review

Implications
The 4 forms of an implication

Which of the other forms of the statement (A -> B) are possibly valid if the statement is valid (Which of the statements, converse, inverse, and contrapositive are valid and true?
Another way to say this is
        If in A then in B.is true or valid

Which of the other forms are valid and true?

Converse case: If in B then in A (B → A).
The converse may not be true and, may not be valid.

Inverse: If not in A then not in B
(~A → ~B). Is this value true? NO

Contrapositive: If not in B then not in A (~B → ~A)
is true and valid!

Many people do not understand implications and the meanings of the four forms.

1.Statement: If you are a teacher, then you like students. If T then you like S. (Assume this statement to be true, see diagram below.)

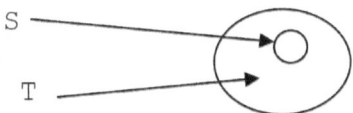

Which of the following are true?

a. **Converse:** If you like students, then you are a teacher.

b. **Inverse:** If you are not a teacher, then you do not like students.

c. **Contrapositive:** If you do not like students, then you are not a teacher. True!

Answers:
   a. This is the converse and may not be true or valid.
   b. This is the inverse and is false and invalid.
   c. This is the contrapositive and is true and valid.

The following figure will help you see the reasoning for the above answers!

People who LIKE STUDENTS
TEACHES Like students

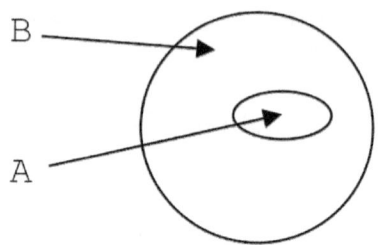

d. How are the following statements related? Valid or not? Or may be

If you live in Montana, then you live in the United States.
Write the other 3 forms and classify each as true or false.

Complete your summary.

Are the family chores, like yard work, house maintenance, getting to be a burden instead of an enjoyment?

Bridge 9: Curves that Control Our Lives!

(You may just want to read this.)

"Analytic Geometry...constitutes the greatest single step ever made in the progress of the exact sciences."

John Stuart Mill

This Bridge will not only review the methods for solving quadratic equations, and investigate their graphs, but will emphasize the applications in our hi-tech world. The more advanced our society becomes, the more mathematical knowledge is needed. Since you are interested in understanding the usefulness of this type of equation reading will be used instead of solving equations. It has been said that these graphs control our society. They are call the conic curves.

Ramp 9.1
Methods for Solving Quadratic Equations

This will be a review for you of some forgotten material. You may just want to read it?

After completing this Bridge, you will hopefully better understand:

1. The definition of a quadratic equation.
2. How to solve a quadratic equation.
3. The definition of a function.

In former Bridges, you solved and worked with equations and their graphs that were straight lines, plus greater than or less than inequations, where the graphs were regions or areas.

The equations had either one or two variables, and were classified as first degree. By first degree equations mathematicians mean that the exponents for the variables are 1. (Since the exponent is 1. We do not write the 1. It is understood to be 1.) In other words, if the exponent is not indicated, we define it to be 1. It has been mentioned that the "game" of mathematics is different from most games. It gets more difficult as it becomes more advanced. The equations you have been solving were(basically) all first degree.

The next type will be second degree equations involving one or two variables and at least one of the terms will have an exponent of 2. Two examples are:

$$y = Ax^2 + Bx + C \text{ or } k = xy.$$

The second example is the case students usually question as a second degree. Good question! Notice the phrase at least one term will have an exponent of 2. The term xy indicates multiplication and when the operation is multiplication the exponents are added, therefore the term xy has a power of two, and that makes it a second degree. It is even more convincing when you graph K = xy. It is a curve, not a straight line.

Bridge 9 will investigate this case.

In a previous Bridges you reviewed the types of equations that were basically y = Ax + B and the graph was a straight line, where A is the slope and B is the y-intercept. In this Bridge the type of equation graphed will be y = Ax² + Bx + C, where and how the A, B and C relates to the graph, plus the methods for solving his type will be investigated. You know, after some thought, what point C is on the graph!

Comment: Notice C is the y-intercept when x equals zero.

Definition 45: A Quadratic equation in two variables, is defined as y = Ax² + Bx + C, where A, B and C are rational numbers and A is not 0.

Function f(T)= 5T + 39 is another example. This could be read as the function T is equal to 5T + 39. The value of the function at 2, written f(2) is 5(2)+ 39 or f(2) equals 49.

Ramp 9.2
Quadratic or 2nd degree equations
A different APPROACH!

At this point the author is electing to take a different approach, which will make the study of quadratics much more interesting. On a sheet of paper draw a 3D cone with the opening upwards and the second cone opening downwards and the two vertically join as shown.

a. Explain how the plane segment can intersect the cone and the intersection is a **circle.** How can this be? Draw this possibility.

b. This time the plane segment moves and the intersection is a **line segment.** How can this be? Draw this possibility.

c. This time the intersection is two **hyperbolas,** one on each cone. Draw the figure to show this.

d. This time the intersection is a **parabola.** Draw the figure.

e. Draw the figure for an **ellipse.**

These intersections, can all be quadratic equations and when graphed **are said to be figures that control our lives.** Think about it! These are used in possibly every profession! A few applications are:

Medical: Focus point for operations.
Communication: Focus point for sound.
Art: All are used!
Games: Billiards, basketball, football, and other games.

The question now is: What is a quadratic equation and how do you solve it, plus is there an easy way to solve it?

To simplify the progress, we will start with the various cases and progress to the solution for all cases! (This is probably a review for you, but may be informational.)

Activity
Curves that control our lives

Graph each equation and name the curve. Then draw the 3D figure.

Problem 1: Y =2x + 1 for values of x an y. Given X = -3,-2.-1, 0,1,2,3. and draw the graph.

Problem 2: Given y = x² and plot the points for x values: -3,-2,-1, 0, 1,2,3. Name the 4q, plane curve and draw the 3D figure.

Problem 3: Given x²+ y² = 1 for x = 5,4,3,2.1. Plot the points, name the curve and draw the 3D figure.

Problem 4: Given x²/16 + y²/9 = 1 for x values: -4,-2,0, 2,4 and sketch the curve, name the figure and draw the 3D figure.

Problem 5: Given x² + Y²= -1 for x values: 5, 4, -2, 0, 2,4. Plot the points and sketch the curve

Can you name applications or point out uses for each of these three-D curves which will help to understand why they are called curves that control our lives. A computer search may help.

Problem 6: $Ax^2 + Bx = 0$
$$x^2 + x = 0$$
$$x(x + 1) = 0$$
(Factoring) x(x+1) = 0

Why are the following answers correct?
Answers: 0, -1

$Ax^2 + C = 0$, will be studied.

Comment: The case where $Ax^2 + C = 0$ leads to what are called imaginary numbers, which will be explained later, if you wish to read it.

The general conclusion is $\pm\sqrt{(C/A)}$ when $Ax^2 + C = 0$.

Comment: Recall that the square root symbol $\sqrt{}$ means the positive square root.

Comment: Notice all second-degree equations have two answers. The two answers may be identical.

If types 1 and 2 don't fit, then try factoring the equation so that the two factors equal zero.

Another type or case: $x^2 - 2x + 1 = 0$
$$(x-1)(x+1) = 0$$
$$x = 1 \text{ or } -1$$

Completing the Square(optional):

If cases 1 and 2 don't fit, then the following method will always work, but as you would expect, it is more difficult.

1. $x^2 + 8x + 1 = 0$. This type cannot be treated as the above types, therefore a new method of solution will have to be used. The following solution was derived over thousands of years.

One way to attack the problem is to try to convert it to one of the above types by adding a number to each side. (In the case above, 15 is added to each side.)

$x^2 + 8x + 1 + 15 = 15$

The equation can now be written as $(x + 4)^2 = 15$ and taking the square root of each side results in: $x + 4 = +\sqrt{15}$ or $-\sqrt{15}$ where $\sqrt{}$ means the positive square root.

———

Solving for x:
$$x = -4 + \sqrt{15} \text{ or } -.13$$
$$\text{(rounded to 2 decimal places)}$$
$$x = -4 - \sqrt{15} \text{ or } -7.87$$

Note: $x^2 + 6x = 2$ (The key is to determine the number to add or subtract to make the left side a perfect square.)
Hint: Try 9

Did you figure out an easy way to determine the number you needed in the C position?

Comment: The number needed is the B/2 term squared, providing A is 1 and you can always operate to make A equal to 1.

This is what you have been waiting for!

The General Formula:

Applying the same procedure to the completing the square method to the general equation ($Ax^2 + Bx + C = 0$)

A general formula can be derived.

   Step 1:
      Given ($Ax^2 + Bx + C = 0$)→
         $x^2 + x(B/A) + B/2A)^2 = 0$

   If uncertain, go back to comment above!

   Step 2:
      $x^2 + x(B/A) + (B/2A)^2 = -C/A + (B/2A)^2$
      The left side now can be written as a square.

      $(x + B/2A)^2 = \quad B^2\sqrt{4A^2 - C/A}$

$$\text{or } B^2\sqrt{4A^2 - 4AC/A^2}$$

Now taking the square root of each side and solving for x, we have:

$$x = + B/2A) \quad \pm \sqrt{(B^2/4A^2 - 4AC/A^2)}$$

Comment: You should simplify the above to the arrive formula.

$$\text{or } x = \frac{-B \pm \sqrt{(B^2 - 4AC)}}{2A}$$

This is a theorem since you proved it and it can be converted to this form:

The sum of the answers is X1 + X2 = -B/A.
(This is an easy way to check answers.)
The product of the answers is C/A.

Theorem 30: If the quadratic equation is $Ax^2 + Bx + C = 0$, then the solution is:

$$x = \frac{-B \pm \sqrt{(B^2 - 4AC)}}{2A}$$

Theorem 31: The sum of the answers to $Ax^2 + Bx + C = 0$ is -B/A and the product of the answers is C/A.

This is a clever way to check your answers.

Comment: You no doubt noticed the formula was written several different ways. This was done intentionally to point out that

$\{-B \pm \sqrt{(B^2 - 4AC)}$ is to be divided by 2A.

This formula will solve all Quadratic equations in one variable.

This is what you have been looking for and it makes the solution quite easy with a calculator. The variable does not have to be x, of course.

Activity

Solve one the following quadratic equations in one variable. When radicals are involved, give the answers in radical form (the exact answer) and then rounded to two decimal places for the approximate answers (use your calculator). The key is to alter the equation to fit one of the types explained above, or use the General Formula.

If you wish, pick one of the following and solve:

1. $x^2-x -6 = 0$
   Sum = -1? Product = 6
2. $3x^2 + 2x - 1= 0$
   Sum = -2/3? Product -1/3
3. $(x/2)^2-x/3 +4 = 0$
   Sum = (-1/3)/(1/4)=?
   Product =(4)/1/4= ?

Investigations
Geometry: A Graph Problem
Optional

1. On a sheet of notebook paper, draw a 6-inch line segment AB about 3 inches from the bottom and parallel to the base edge of the paper. Then mark a point one inch above the line near the center of the sheet. Label it P. Now the problem: Locate several points that are just as far from the point P as each point is from the line AB.

Hint: The easiest point to locate is the midpoint of the perpendicular from O to line AB. Now locate a few points!

Connect the points to form a curve.
    a. Describe the curve.
    b. Draw a line from P to the point (C) and reflect it so that if it were continued to the segment AB it would b perpendicular to AB.
    c. If this curve were rotated it could be an example of an auto head light. The rays are all parallel and provide light on the road.

Comment: Look up the word parabola" in the dictionary.

2. Using graph paper construct the graph for $Y = X^2$ using the x (integers) values – 5 to +5 include 0.

Ramp 9.3
What graphs are!

"The essence of plane Analytic Geometry is in the matching of ordered pairs of real numbers with the points on a plane."

Edna E. Kramer

This section could be very time consuming and since the objective is for you to get the general picture, so select a few problems from each and observe the conclusion. After completing this Ramp you will hopefully be able to:

1. Visualize the graphs for the quadratic equation. $Y = 2X^2$ these points for x, –3 to 3. Plot the points and sketch the curve.
    a. Where do you think the turning is?

b. Draw an axis of symmetry.

The graphs are more interesting and more meaningful plus interesting if a computer or graphing calculator is available.

Another Conic section

2. Drawn the graph for $x^2 + y^2 = 16$, for x from -4 to 4.
   a. a. Plot the points, sketch the curve.
   b. b. Name the curve. Does it have an axis of symmetry?
3. Draw the graph for $x^2 - y^2 = -16$ for values of x from -4 to $+4$.
   a. a. Sketch the curve.
   b. b. Name the curve.

Graphing is the gift of Rene Descartes.

Ramp 9.4
ACTIVITY

Some interpretations and Application and conclusions!

Property of ice

1. Fill a glass half full of water.
   a. What are the three possibilities for an ice cube when you drop it in the water? Hint: One possibility is the cube will not float, it will float, it will sink to the bottom.
   b. Now put in an ice cube, which possibility is the correct one?
   c. What does this mean about the ice cube?

d. What would be the consequence in lakes and rivers be, if the ice cube had sunk to the bottom of the glass?

Answers:
a. Sink to bottom, float on the water, partially submerge.
b. Float
c. The weight of the cube is less than the weight of the volume of water displaced. Water expands as it freezes.
d. Ice would settle to the bottom of the lake and rivers resulting in the death of aquatic like.

Thinking: Taxicab geometry

2. Taxicab geometry is different from Euclidian Geometry. One difference is in the negation of the "shortest distance between two points is a line segment."

On the grid (like the streets of a city) below, answer the following questions.

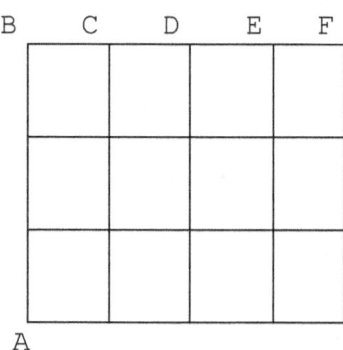

How many ways or routes can a taxicab driver travel from A to C using the shortest distance requirement? The taxi drive must always go to **the right or up** on the streets.

a. How many ways can a taxicab driver go from A to D

b. Complete the following table to show your answers.

Route or Path	Ways
A->B	3
A->C	?
A->D	?
A to E	

Try to predict the number of ways go in from A to E?, then A to F? Predict your answer first, then count the ways. (This is Inductive reasoning.)

Answer: 6, 18.

You may invent clever ways to predict the answers.

Ramp 9.5
What Is The FOCUS POINT?
With Applications

Three-dimensional parabola

If you studied each of the above applications you would notice in each case there is a "point" which corresponds to the F point in the class activity.

This point, called the FOCUS point, and is very important in order for the applications of the parabola to function. It will be our objective in this section to be able to locate the focus point given the quadratic equation for a parabola.

The interesting and useful properties of the parabola are:

1. If light or sound waves are generated from the focus point, then the waves are parallel after reflecting off the parabola.

2. If parallel incoming waves hit the parabola, then the reflected rays will intersect at the focus point.

In order to understand this useful curve, you need to:

1. Understand the algebraic and geometric definitions of the PARABOLA.
2. Be able to derive the coordinates
3. of the focus point given the quadratic equation of the form $y = Ax^2$
4. Recognize some of the applications of a parabola.

The above, no doubt will create a question in your mind as to the algebraic definition or equation for the parabola. The distance formula from geometry, which is really the Pythagorean Theorem, will help you derive the equation.

Theorem 33:
The distance between two points is:

$$D = \sqrt{(x-x_1)^2 + (y - y_1)^2},$$
where x,y) and (x_1,y_1) are the two points, and D is the length of the line segment. This really is the Pythagorean Theorem.

Activity

Now you are ready to solve for the equation of a parabola and in turn the coordinates of the focus point. The following figure will be used. Given the equation $Y = X^2$ and we want the focus point and the directrix. The rule is that any point on the curve is half the distance to the turning point as the point is from the directrix.

Graph the equation y = x² for the x values -2,-1, 0,1,2 x and sketch the curve. Label the points A, B, C, F, and E in the same order as the x values. The focus point (0,1) is on the y axis, the line of symmetry directrix is the line y = -1. Plot the curve, fill in the points, and also the focus point and the directrix. Draw the rectangle with corner points A(0,4), B(2.4), C(2,1), F(0,1) and segment BF is the reflected ray to the focus point. The value for focus point is the distance the focus point is above the turning point or in this case the x-axis.

Graphing
(Be Selective)

2.The turning point is also identified as the **maximum** or **minimum** point depending on which way the graph opens. The graph y = x² opens upward and the point (0,0) is called a minimum point. If the graph opens downward, then the turning point is a maximum point. This is determined by the value of A in the equation type y = Ax². Maximum and minimum are very important in industrial problems, concerning income and expenses.

Ramp 9.6
The Conic Sections

"It can be of no practical use to know that π is irrational(number), but if we can know, it surely would be intolerable not to know."

E. C. Titchmarsh

Applications of the Parabola
(Basically for only reading)

The objective of this section is to give you more practice using the concepts and skills you worked with in former Bridges. The objective will be to help you understand more applications for the parabola or the properties of the paraboloid, which is the name for the three-dimensional parabola.

The parabola has many applications, a few have been mentioned in the previous sections. Hopefully, you will recognize additional cases when you see them.

The following is a list of items that illustrate the applications of the parabola:

- Satellite dish
- Automobile headlight
- Searchlights
- Loud speakers
- Cell tower receivers and transmitters
- Path of an object thrown in the air
- Path of a bomb dropped from a plane
- flashlight reflectors
- Some types of optical lenses
- Suspension bridges
- Radar
- Reflecting telescopes
- Some microphones

Comment: Create a set of photographs showing the applications of parabolas in your community.

Research:

1. The uses or applications of conic sections
2. Mathematician: Appollonius
3. Draw two lines that intersect and then rotate them to form a 3D figure.

Investigations
Optional but Interesting

Geometry: Locus (What is your definition of locus?)

In the approximate center of a sheet of paper, mark two points 8 centimeters apart and label them A and B. Draw segment AB. Locate 10 points (P) that meet the following condition: AP + BP equals 10 centimeters.

Comment as to the figure. Connect the points with a curved segment.

Bridge 10: Imaginary numbers

"The definition of a good mathematical problem is the mathematics it generates rather than the problem itself."

Andrew Wiles

COMMENT:
Dr Wiles was honored in the 1990s for PROVING Fermat's last theorem. I think he received $700,000. Do a computer search for Fermat.

Ramp 10.1
Complex Numbers Plane
Read only

These are the types of real numbers:

1. Counting Numbers
2. Whole numbers
3. Integers
4. Rational numbers
5. Irrational numbers

Complex numbers are an addition to these real numbers on the number line where the Complex numbers are all the numbers on the number plane. This is the geometric way to describe these complex numbers

Research: You may wish to prepare a report on the history of imaginary numbers. Rafael Bombelli (1530+), Casper Wessel(1745-1818), and others will provide some leads.

These complex numbers are used in the fields of electronics involving radar, x-ray, radio transmissions, television and many medical applications, etc. You should be realizing why engineers are required to study so much mathematics. Also, why many of our everyday conveniences or luxuries, which we take for granted, depend on mathematics.

RAMP 10.2
The Sweet Spot
Interesting Applications

A batter tries to hit the ball on what is called the sweet spot. This is really the center of gravity and transmits to the ball the maximum energy and consequently, the ball will travel the farthest, and hopefully will be a home run. The question is how to locate this point? There are several ways, ask your coaches. One way is to lay the bat on the floor or a flat surface and drop a golf ball or better a ping pong ball on the bat starting at the end of the bat and moving the drop point about 1 inch at a time up to the handle. Drop the ball from a height of 2 to 3 feet, but be consistent. The point at which the ball rebounds the highest is the sweet point. This point is also the center of gravity.

WRITE A SUMMARY

Bridge 11: Conic Sections Influence Our Life

"We learn the new in the light of the old."

Anonymous

Ramp 11.1
Conic Sections
Activity: Read only

Using graph paper plot the following points which satisfy the equation:

$$x^2 + y^2 = 25$$

1. Complete the following and plot the points on Graph paper. Plot the points from -5 to -5 for X. If the number of points were increased and connected, what do you think the curve is? Name it.

The following locus definition for a circle is from your geometry class.

Definition: A set of points on a plane that are equal distance from a given point (center)on a plane is a circle.

The following questions or terms are to refresh your memory as to the definitions or formulas relative to a circle.

   a. The point referred to in the locus definition is called the center.

    b. What is the name given to the "equal distance"
    phrase? Answer: Radius (pl. radii).

Do you recall the definitions of the following,
related to a circle?

    a. Diameter
    b. Circumference
    c. Area of a circle
    d. Chord of a circle
    e. Tangent to a circle
    f. Secant to a circle

2. What is the formula for the circumference of a
circle?

3. What is the formula for the area of a circle?

4. In the locus definition above, why is "on a
plane" necessary?

    Answer: To distinguish it from a Sphere.

5. Now you have will two descriptions or definitions
for a circle! What are they?

6. The algebraic formula for a circle can be justified
from the following figure where the point (x,y) is
any point distance from the point (0,0). The formula
for a circle in algebra is: $\mathbf{x^2 + y^2 = r^2}$, where r is
the radius and the center is given.

Note: A circle is the set of points (x,y) such that
$x^2 + y^2 = r^2$.

Your question now may be: What if the center is not
at the origin, then what is the equation for the

circle? Take the following case, where the center is at (a,b) and r is the radius or distance.

Recalling the distance formula or the Pythagorean Theorem, the following is true for the general case?

$$(x-a)^2 + (y-b)^2 = r^2$$

Comment: Yes, by applying the Pythagorean Theorem. This is also an algebraic formula for a circle and will be stated as a theorem. Circle = $(x-a)^2 + (y-b)^2 = r^2$, where r is the radius and the center is at the (a,b).

The problem early mathematicians had was this. Construct a circle given the circumference or the area. The constant **K** was not given the name until the 1600's by Euler and **π** was not proven until the 1800's (by Beckmann).

Interesting Research Project
Diagonal of a 4th dimensional cube

What is a tessaract?

Ramp 11.2
The Ellipse
(Read only)

The important thing is not stop questioning.
Albert Einstein

An ellipse looks something like a racetrack oval or the running track usually surrounding a school's football field. It is also a very important curve used in industrial areas, space navigation and recently in the medical field. The orbits of planets are elliptical.

After you study this Bridge, you will be able to:

1. Define an ellipse
2. To graph the equation for an ellipse.
3. To derive the equation of an ellipse, given the necessary information.

In a previous Bridge, you were asked to envision how a geometric plane could intersect a cone to form a circle.

This is certainly easier for you to visualize now. Can you imagine how a geometric plane segment can intersect a cone to form a figure similar to a racetrack, called an ellipse?

Comment: A geometric cone can usually be purchased at a crafts store and you can slice it to form an elliptical cross section.

The ancient Greeks observed this curve by passing a plane through a cone just like you did in a previous Bridge. Are there any limitations placed on the angle for the beam to the plane?

The geometric definition of an ellipse is the set of all points P on a plane such that given two points F_1 and F_2, then the set of points P that meet the condition $PF_1 + PF_2 = K/2$. (K is a positive number equal to the major diameter.)

Picture-wise the above definition is shown, for one point on the ellipse, in the following diagram. Y

$$PF_1 + PF_2 = K/2$$

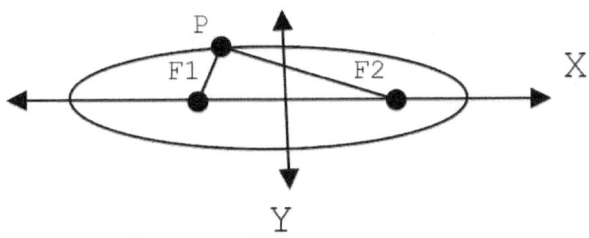

(P is any point on the curve of the ellipse. Hint: My use the 3, 4, 5 right triangle. K is the major diameter between the focal points, F1 and F2.)

Activity

1. On a plain sheet of paper, mark two points, F_1 and F_2, that are 8 cm apart. Now locate 10 points that satisfy the condition, $PF_1 + PF_2$ equals 10 cm. Two of the points are to be on the line determined by F_1 and F_2 and two are to be on the perpendicular bisector of the segment F_1 and F_2. This is a good project for a math teacher to assist you in locating a few points and reflect the points over the F_1 and F_2 line. Then sketch the curve through the a few points. The figure is an ellipse and should be similar to the following.

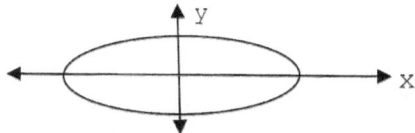

Complete the following table using the equation $x^2/25 + y^2/9 = 1$ and plot the points on graph paper for X from −5 to 5.

   a. Now plot the points and draw a curve containing them.

   b. Locate the focus points and the length of the major diameter.

This leads to the algebraic formula for an ellipse.

The general formula for an ellipse is

$$x^2/a^2 + y^2/b^2 = 1$$

1. In this formula, the values for "a" are the x-intercepts. The values for "b" are the y-intercepts. The formula for the focus points is:

$$F = \pm\sqrt{(a^2 - b^2)}$$

F is the distance from the origin.

Activity

$$x^2/5^2 + y^2/3^2 = 1$$

With 6 points plotted, a fairly good sketch of the ellipse can be drawn.

The following is a summary for your notes.

If given the formula for an ellipse
    $x^2/5^2 + y^2/3^2 = 1$, then:
  X intercepts are (-5,0) and (5,0)
  y intercepts are (0,3) and (0,-3).
  The focus points are (4,0) and -4,0)

The **physical property** of an ellipse is important. This unique property is: If you start an object like a ball or a ray from the focus point F1, then the ball or ray after hitting the side will pass through F2 The above case is a major property of an ellipse The major axis is the segment (5,0) to

(-5,0) and the minor axis in the above case is (0,3) to (0,-3).therefore if you start an segment from F1 and it hits (0,3) then it rebounds to F2 and it hits (0,3.Draw the figure and what kind of triangle is improved? Therefore, what kind of tiangle is formed and what is the length of (0;0) to F1?

Interesting Uses of Ellipses

Some museums have what are called a whispering room. The ceiling of the room is in the shape of a section of an ellipsoid and two people each standing at the focus points can whisper and hear each other, but no one else in the room and hear them.

In recent years this principle of the ellipse has been used in the medical world to smash kidney stones using sound waves by placing the sound generator at one focus point and the kidney stones at the other focus. The patient is placed in a semi-ellipsoid shaped tub filled with water. The sound waves travel through the water and the body to the kidney stones and the frequency or vibrations break up the stones. (The author had this operation.)

Comment: Ask a doctor or nurse to explain this application and the procedure. The story behind this discovery is interesting involving aerodynamics. Two Army pilots and their wives, who were nurses came up with idea.

Another application is the earth's path in space. The center of the sun is one focus point for the earth's elliptical orbit around the sun.

Another way to draw an ellipse

1. On a peg board or similar material (card board) locate two points and to each attach the ends of a piece of string (longer than the distance between the two points).

Now using a pen or pencil trace a curve by stretching the string with the pen and moving the pen plus marking the curve.

Ramp 11.3
The Hyperbola

"Relationships between different subjects
Even BRANCHES of Mathematics are
creatively important in mathematics."

Simon Singh
FERMAT'S ENIGMA

The following activity will help you mentally see a hyperbola before we formally define the curve. Given the following figure: Do you see the figure the geometric plane makes when it intersects the cone?

Sketch what you think the intersection will look like?

Comment: The Intersection is the two branches of the hyperbola. Be sure to continue to envision the intersection until you "see" the two branches and the condition that the plane must not be parallel to one of the edges of the cone.

This is another conic section, and is related to many everyday problems or activities. A few are:

1. The force necessary to keep a car on the road is hyperbolically related to the radius of the curve.
2. The illumination from a light is hyperbolically Related to the distance from the light source.
3. The gas pressure in a tank is hyperbolically related to the volume of the tank.
4. The electric current in a circuit is hyperbolically related to the resistance of the wire.
5. The support two parallel beams will provide is hyperbolically related to the distance they are apart.
6. The number of vibrations a guitar string makes is hyperbolically related to the diameter of the string.

Certainly, the above statements all depend on the meaning of the word "hyperbolically." This Ramp will explain this concept in more detail.

You should be coming aware of the importance of the conic sections in our world and why some people label the conic sections as the "curves that control our lives." The conic sections we have studied so far are the circle, parabola and the ellipse. Apollonius named the curves in these Bridges as Conic Sections since the visualized them by the intersection of a plane and a cone. Which date do you think indicates his time period, 1940, 1776, 1492, 500, 225 B.C.E?

Research:
1. TIME magazine, Feb. 23, 1959, p. 87 for information on DECCA (air navigation) involving the hyperbola.
2. The geometric or locus definition for a hyperbola is: Given two points (focus points or foci) the set of points which meet the following condition is a hyperbola, $F_1P - F_2P = K$. This means the distance

any point P is from one focus point minus the distance the point is from the other focus point is a constant.

Activity

On a sheet of graph paper, plot the points, which result from the possible answers to the following area problem.

The following example will show you the easy way to construct the graph for a hyperbola given the general formula,

$$x^2/a^2 - y^2/b^2 = 1.$$

Notice, the curve for equation in the form, $x^2/a^2 - y^2/b^2 = 1$, has one curve in quadrants 1 and 4, and the other curve in quadrants 2 and 3.

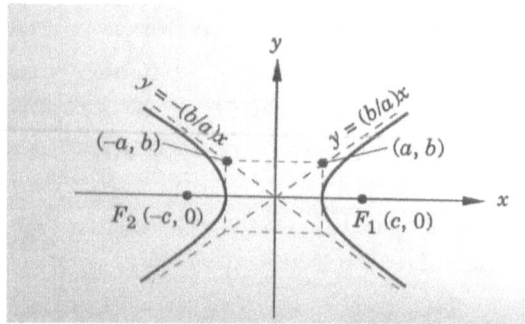

What are the coordinates for the four points on the rectangle? Use the above graph to show that K equals 2a.

$$\text{D}\{F_2 \text{ to } (a,o)\} - \text{D}\{F_1 \text{ to } (a,o)\} = 2a = k$$
(The above was a handout at a NCTM meeting.)

The lines y = bx/a and y = -bx/a are called the asymptotes. What do you think the definition of asymptote is, judging from the figure?

Comment: Asymptote is from the Greek and defined as "not meeting." In this case, the hyperbola approaches the line (diagonals) but never intersects it.

Complete your notes:

> **"Mathematics consists of islands of**
> **knowledge in a sea of ignorance."**
>
> Simon Singh
> FERMAT'S ENIGMA

Topic for dinner and/or after dinner could be retirement incomes or plans. When to sell the house and why?

The next diagram enables you to see the conic sections as a plane intersects with a double cone. Good for just reading enjoyment.

Can you draw the 5 conic intersections of the plane and the 2 opposite cones?

There is a paperback book by Meyer titled FUN WITH MATHEMATICS (See bibliography) which contains a chapter called "Curves That Control Our Lives." This is interesting reading and is recommended. If you or children have a hobby of photography including video, perhaps a photo display or video show could be created to illustrate the use of conic sections in your community. Eventually you will see "conics" everywhere from the wheels of an automobile, the conical shape of a pine tree, the cross section of a cylindrical water tower, the parabolic arc of a punted football, to the cone filled with ice cream on a hot day. Can you identify the conics in the figure that you just drew? Can the plane intersection ever be a point or two straight lines? (Yes, explain.)

You may find the biographical sketches of the following mathematicians interesting and informative:

Charles Babbage,
Geo Boole,
Jon Von Neumann,
Howard Aiken,
L. Euler,
B. Pierce.

These individuals contributed to scientific advancements as indicated:

- Babbage (computer concept),
- Boole (symbolic algebra-basic for computer),
- Von Neumann (stored program concept),
- Aiken (electric computer-Mark I),
- Euler (Many areas) including the use of the symbol for Pi

Bridge 12: Archimedes, Logs, Exponents

Read Only

Number Investigations
Archimedes Theorem

Old Math: Archimedes discovered an interesting relationship between the cylinder, sphere and the cone when the sphere and cone are inscribed in the cylinder (The radius and the altitude are the same for all three figures.

1. Draw the figure:
   a. Draw the picture showing the sphere and cone inscribed in the cylinder.
   b. R is the radius of for all three figures? What is the altitude of each?
   c. What is the formula for the volume of each? See index. (Leave the answers in terms of pi.
   d. What are the following ratios?
      1. Vol.(sphere to Vol. of cylinder) in the same cylinder
      2. Vol.(volume of sphere to Vol of cone in same sphere)
      3. Vol. of cone to vol. of sphere.2. (1600 BC)

Time for a few Interesting Review Problems
(Be selective)

2. Given a rectangular box with dimensions (LWH) 12 inches by 18 inches by 15 inches.

a. Using your ruler draw a box to scale where 1 inch equals 6 inches.
b. What is the surface area, include the top?
c. What is the volume.
d. What is the length of the diagonal?
e. Could this box be shipped via the U.S. postal Service? (What are the U.S.P.S. regulations? Check with

3. Who do you think exerts more pressure per square inch on a football player while standing on only one leg, a 250 pound football player with 2.5 inch square heels or a 120 pound cheerleader wearing .5 inch square heels standing on one foot? Guess first and then calculate the pressure per square inch for each.

Answer: The football player exerts 40 pounds per square inch and the cheerleader exert 480 pounds per sq.in.

Thinking Problem

4. How could the flag be designed to keep the rectangular shape if there were 53 States. Draw this rectangle.

Answer: 8787878

5. What is your answer to each of the following?

a. What is the probability of tossing a 2 with a die?
b. 5 meters = how many feet?
c. 4 pictures can be arranged how many ways?
d. What is the probability of throwing a pair of 2s with a pair of dice?

6. At 6% interest per year, how many years will it take for the money to double?

7. What is the graph of y=mx + b, and were does cross the y and x axis?

8. John has an investment at the local Savings and Loan for 7 years, The original investment was $1000 and now 7 years, the value is 1800. What was the rate of interest to the nearest integer?

Ramp 12.1
Some History, EXPONENTS

The idea of fractional exponents was conceived by the Bishop of Normandy, Nicole Oresme, about 1323-1382, and re-discovered or re-invented by Simon Stevin, 1548-1620, who is credited for our way of writing exponents. (Check a history of mathematics source in the bibliography.)

What is 25 to the ½ power or exponent. What is the answer.

Age old question: Is mathematics discovered or invented?

Application: National Debt

1. If the National Debt is approximately $6.75 trillion, then:

   a. Write the above figure using $675 and zeros.
   b. Write the debt figure in scientific notation.
   c. What would every man, woman, and child in the U.S. have to contribute in order to pay off the debt? Assume the population is 300 million.

---

d. What is the current U.S. debt? (Search the Internet for the answer.)

Read Only!

(I don't think Logs are taught in the high schools today! Invite a math teacher to give a talk on Logarithms'

"The invention of logarithms Laplace said amounted to "shortening the labor's (and) doubled the life of the astronomer."

F. Cajori

Application: Sound

1. Which medium will transmit sound travel better, air or a solid wall? Select your answer, then conduct the following experiment with a friend.

   a. Stand at one end of a room and your friend at the other end of the same wall. Tap the wall with your fingernail or a pencil. Can your friend hear the tapping?
   b. Now have your friend put his/her ear to the wall and your tap the wall again. Can the other person hear the tapping?
   c. Repeat the experiment until a conclusion is reached.

      Answers:
      a. No. Should be yes.
      b. Denser material sounds better.

Thinking: Easy Money

2. You and your friend are having a coke after school in the lounge. You put a half dollar on the

table and your friends does the same. The same process is repeated two more times. How much is now on the table? Your friend now offers you the coins on the table if you will give her $2. Should you accept this "good" deal? Explain or justify your answer.

> Answer: Do not accept the "good" deal. Act it out and you will see why.

Ramp 12.2
Application: Heat transfer

3. There are three ways that heat is transferred. The following activity will illustrate the three ways. Your problem is to list or summarize the ways.

   a. Light a candle and hold your hand about a foot above the flame. (Remove your hand if it gets too hot.)
   b. Light a candle and hold your hand about 4 inches to the right of the flame and gradually move your hand closer to the flame until you feel the heat.
   c. Light a candle and while holding one end of a key, or a spoon, put the other end in the flame. Hold the spoon in that position until you begin to feel the spoon getting warm.
   d. Summarize the three ways heat is transferred. Check an encyclopedia for the explanation why a thermos bottle will keep a liquid hot in the winter, or cold in the summer.

> Answer: Heat transfers by convection, conduction, and radiation.

4. How does a 7 rated earthquake compare with one rated 8? Application of Logs!

(If a Log Earthquake is rated 7 and earthquake E is 8 then the E is 10 times larger or dangerous than R.

Bridge 13: Course Review

Ramp 13.1
Inductive Conclusions
(A Must)

Note: This Bridge can be worked as a group project so all will benefit from it, You know what inductive reasoning is, and the weaknesses involved in the resulting conclusion. This Bridge will provide you the opportunity to review these ideas and add a few more activities and methods and gain from the understanding of others. Even invite a math teacher to assist you.

Someone once stated the following, probably a Math teacher.

"We learn the new in the light of the old."

Inductive Reasoning and a few interesting problems

In a former Bridge you studied inductive reasoning and concluded it had a weakness. What is the weakness? If you recall, inductive reasoning was defined as a method of reaching a general conclusion by observing a few cases, and then assuming the rest of the future cases followed the same pattern. Another way to simply state a conjecture and with the help of your teacher or mentor you proved the conjecture. Once the conjecture was proven, you classified it as a theorem. Remember! the concept, would be that a conjecture is reached from observing a few cases. In your geometry course, you probably drew a few geometric figures, then stated a conclusion. This type of reasoning is used every day by everybody!

This section will be using inductive reasoning to make conjectures or conclusions and emphasizes they are based on assumptions: Some cases are listed below for you to practice making conjectures.

Activity
(Be selective)

Your conjectures may come up with different answers than are expected. This makes for an interesting activity.

1. a. $2 + 4 =$
   b. $4 + 8 =$
   c. $6 + 12 = 18$
   d $8 + 16 = ?$
   e. $10 + 20 = ?$.

Write 2 conclusions for the sum of the next pair of even counting numbers or any consecutive pair of even consecutive even pair.

2. Write a conjecture for the sum of the first N counting $(1+2=?, 1+2+3=?, 1+2+3+4+=?)$ and so forth.

3. General case? Sum is ?. If you are interested in the proof invite a math teacher to your group, but tell him or her the question first so he can be prepared!

Comment: On problems of this type it is a great aid to make a table as below to help "see" the solution.

N (cases)	The sum of the first n digits	
1	1+2	3
2	1+2+3	6
3	1+2+3+4	10
4	1+2+3+4 5	15

From the above, can you predict the sum for the following?

1+2+3+4+5+...N = ?

5.(A must!) After observing a few simple cases, write a conjecture for the sum of the first set of counting numbers, but it is an assumption, due see why?

6. After observing a few simple cases, write a conjecture for squaring counting numbers ending in 5.

  Case 1.   5 squared is 25
           2.   15 " is ?
           3.   25 " is ?
           4.   35 " is ?

Look for a pattern!

Answer: $(C5)^2 = (C+1)^2$ and add 25 where c,c+1 and ad 25 Example: when C is 25: 3 times 2 and add 25 = 625

What is 55 squared? Answer is 30 and add the 2 digits number (25). The answer is 3025. work a few cases with your friends!

7. The "divisible by nine" conjecture
      a. Write a few numbers which are divisible by nine.

b. Add the digits for each number in part
a. If the first addition is a two-digit
number, add again.) Example: --> 18 --> 9
c. Write a conjecture.
d. Notice your conclusion is not proved and is
valid for your few cases, unless you prove
it for all cases.

Comment: This is called Casting out Nines. Write
the assumption!

8. PROBLEM (A good group activity.)

a. Draw a circle with 2 points on the circle.
Connect the points and now you have
b. 2 regions with-in the circlE
c. Now add another point on the circle and put
in the segments, put in the segments and now
you have how many? segments, points ? and how
many regions?
d. Keep adding points and the number and check
your predictions. List your predictions. Check
your answers by counting the numbers for
Points? Segments? Regions? Check your answer
by counting Regions.
e. Draw the circle and put in the points and
check the numbers with
f. 5 vertices, the ? of segments and 5 regions.

Euler concluded: vertices + regions = segments
+2 and he proved his predictable answers

Geometry: Golden Rectangle and/or the Golden
Ratio

9. In the professional field of architecture the following figure is known as the Golden Rectangle.

This rectangle is said to be the most beautiful (or pleasing to the eye) rectangle. (Ask an art teacher to visit your family one evening.)

  a. Draw a rectangle which in your opinion would be the ideal size for your art piece.
  b. What is the ratio of L to W to the nearest tenth of an inch?

Now let's calculate the ratio the ancient artists said the formula to derive the measures for the sides is $L/W = W/L-W$. This an equation with 2 valuables so let us substitute 1 for L so we have only one valuable to solve for W.

To solve for the measurement ratios, let L be 1 and solve the equation for W.

  a. To actually determine the ratios, assume the length is 1 unit and solve for the width.
  b. Draw, to scale, the rectangle determined in "a."
  c. An interesting property of these two numbers is that they are reciprocals of each other. Check this property with your calculator/computer.

Answers:
  a. 0.618 or (−1+.5)/2
  b. 1/.618 Draw the rectangle that is the size of the most beautiful rectangle,1 is the length and 0.618 is the width.

Comment: For additional information pertaining to the Golden Rectangle, try a computer search or ask an art teacher!

Application: Baseball Bats

10. Which bat is better?

  a. Which bat transfers more energy to the ball, or a lighter bat, which you can swing faster or a heavier bat with a slower swing? The formula for energy transfer is E(energy) = (1/2)M (mass)times V^2(velocity squared)or ask the coach.
  b. If the mass is doubled, then what happens to the amount of energy
  c. If the velocity is doubled, then what happens to the amount of energy?
  d. If the mass is reduced to 3/4 the original mass and the velocity is 5/4 the original velocity, then how has the energy changed?
  e. If the mass is increased to 5/4 the original mass and the new velocity is 3/4 the original velocity, then how has the energy changed?
  f. The energy is transferred to the ball upon impact, then which bat is preferred, a lighter or heavier one?
  g. Write your answers and explanations

Answers:

  a. Energy varies as the square of the velocity. (Lighter bat)
  b. Energy is doubled
  c. Energy is quadrupled or 4 times as much.
  d. Energy is quadrupled or 4 times as much.
  e. 1.17 times the original energy.
  f. The new energy is .70 times the old energy.

    g. Devise a way to test your answers, but you must hit the ball on the "sweet spot" or center of gravity

Comment: Ask the local coach to demonstrate how to locate the sweet spot.

The first set of problems all related to math, but to be a better decision maker you must recognize the use of inductive reasoning in your everyday world. A few old sayings are:

    a. Friday the 13th is bad luck day.
    b. It is bad luck, if a black cat crosses your path.
    c. Break a mirror means 7 years of bad luck.
    d. Ask your friends for a few more inductive sayings.

11. Many of our activities are taken up because of inductive reasoning and change with a person's application of inductive reasoning. If we like an exercise, then we may continue the exercise with or without the recommendation of a trained physical therapist. We play games that are fun because it feels good whether or not it is shown scientifically to increase hormonal response in the body. This is a form of inductive reasoning. Practices like Tai Chi or yoga gives exercise and is also meditative, it feels good and is healthy, so then I continue the practice because it is worthwhile to be healthy. Throughout our lives, we play sports games, card games, dance, and tell stories because these activities are fun. Doing things because we have fun and enjoyment is justified by our own inductive reasoning.

My son has done Tai Chi for over 30 years \ because it continues to give him many benefits. He also does an adult form of improvisational movement and story telling, called InterPlay, that he enjoys immensely. Check it out at InterPlay.org!

Ramp 13.2
Inductive Conclusion and its Weaknesses
(A must!)

Rodin's The Thinker

INDUCTIVE REASONING ACTIVITIES

"The important thing is to not stop questioning"
Albert Einstein

In this Bridge you will review inductive reasoning and some of its weaknesses in non-mathematical situations. The weakness in inductive reasoning is in the conclusion assumption or conjecture, which is a statement about the future made from a few cases. This was and is a weakness used by many people in decision making and is why geometry (in theory) is required in many high schools for graduation. Superstitions are consequences resulting from inductive reasoning usually from observing a few cases. We naturally draw conclusions from the cases, but we must realize the weakness of predicting from these few cases. Many the conclusions are from visual cases and what we think we see is not always, but most cases the non-mathematical situations are not used. As Prof. Fawcett justified in the NCTM 13th Yearbook, Nature of Proof, if critical thinking is to transfer to everyday situations, then those types of situations must be analyzed and covered in the geometry course.

Examples: The following has been sent to the author.

A few weaknesses

1. In the following diagram, do you see 3 or 5 boxes? Most see only 3, but when the figure is inverted picture, you may see 5.

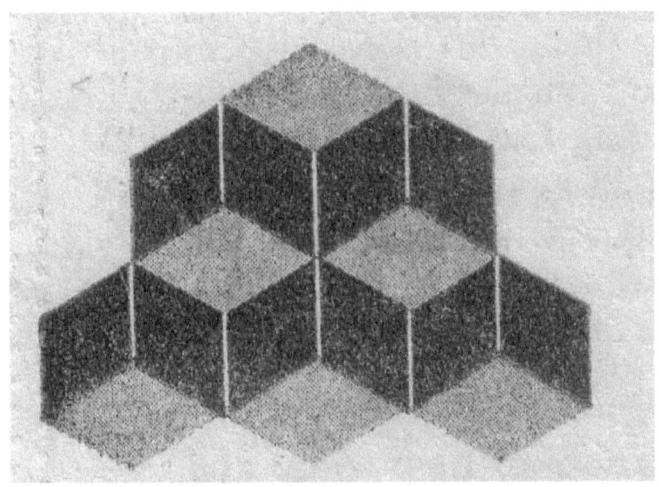

Keep looking until you see 3 or 5. The figure appears to have the small box in the corner, but by moving your head to the right or left of the box you may suddenly see the other case.

2. Where is the small box, in the corner of the box or outside the big box?

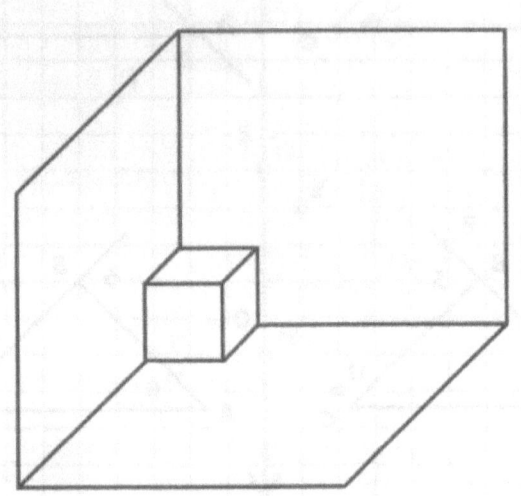

3. Look in the encyclopedia for a picture of the Canadian flag. Can you see the profiles of the faces of two old men in the flag? (Focus on the white color.)

Ramp 13.3
A Family game called Sprouts
Optional

This game of Sprouts* is for two players. This game was invented by J. H. Conway and M. S. Paterson in 1967 and may take some time to learn and play. The objective is to be able to predict (1) who will win, (2) the player who makes the first move or the player who makes the second move. The third objective is to predict the number of moves to complete the game given the number of sprouts to begin with.

Rules and definitions

Draw a half inch circle to represent a Sprout. A MOVE is drawing an arc from a sprout to another sprout, or to the same sprout and creating a new sprout at the arcs midpoint. A sprout is dead (not usable when it has 3 arcs.)

1. The game is over when there are no possible moves. The last player to make a move is the winner.
2. A legitimate sprout is any sprout which has only one or two arcs.
3. A sprout with 3 arcs is dead and out of play.

A simple game to illustrate the rules will start with one sprout, Sprout A with one point.

Move 1: Player#1 will draw an arc from A back to A and mark B the center of the arc. Notice A and B each have now two arcs.

Move 2: Player 2 now marks C as the midpoint of a new arc from B to A. Which sprouts are now dead?

Move 3: Player can now draw an arc from sprout A to B and mark the midpoint D the midpoint of the arc C. Three arcs from a sprout and the sprout is dead or out of play.

Now practice!

The next game starts with 2 sprouts.

The **winner** of the game is the player who makes the last possible move. Record the number moves and who won.

Result from first game is: Player A has 2 moves, Payer B has 1 move and is the winner. There are a total of 3 moves.

Notice: The moves continue until there are no more possibilities. This is why the number of moves by each player must be recorded. There is a pattern that will enable you to tell who will win and in how many moves if you know the number of sprouts to begin with. Now play the game with 2 sprouts and 2 players, then with 3 points and 2 players. (Read the rules and the objectives again. Have fun!)

Objective: If the 2 person game starts with n points, then the objective is to be able to predict the number of moves it takes and who (by induction) will win?

Comment: The game can be played with three or more persons, but the generalizations will be different.

Suggestion: Play this game with your friends and they will be mystified, since you may know who will win and the number of moves it takes to win.

Ramp 13.4
Interesting Geometric Relations
(Be Selective!)

1. Geometry: **Euler's Formula**
   Euler's formula can be adapted to plane figures or three- dimensional figures. We will first apply it plane figures.

   a. A simple plane figure could be a line segment consisting of points A and B (the end points) and the segment, plus the number of sections or regions the plane is divided into.

   b.

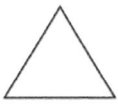 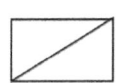

Figure 1   Figure 2
  #1   #2

The number of vertices is ?

Answer: Figure 1: 3, Figure 2: 4

The number of segments is ?

Answer: Figure 1: 3, Figure 2: 5

The number of regions is ?

Answer: Figure 1: 2, Figure 2: 3

If you would like an interesting two problems try to solve for the number of segments related to the number of points (like in number 1 above and the regions as in number 2 above.)

Answers:	Points	Segments	Regions
	2	1	2
	3	3	4
Objective	N	?	?

From the above two cases, you may predict the next set of answers and continue until you get to the answers for at least 7 points. You will be surprised!

Another interesting case

Euler's formula, V + R = S + Continue until you see the pattern in each case and you can predict the answers.

The formula for three-dimensional figures uses the vertices, segments, and faces (F). A face is defined as a plane segment like the top, bottom or sides. Count the hidden points, faces and segments also!

Euler's formula for a cube:
V + F = S + 2

In a cube there are:
    number of vertices is ?.
    number of segments is ?.
    number of faces is ?.
Is **V + S = R + 2** valid?

Test the formula using this pyramid a 3-D figure.

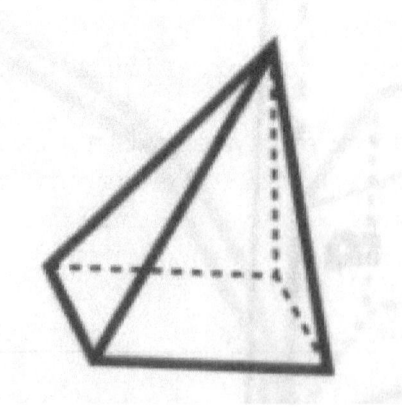

Vertices + regions = sides
5 + 2 = 5 + 2

Draw a few more 2D and 3D figures and test the formula. Does V+R(regions) = Sides + 2?

Your answer should be yes.

Application: Reflections

Mirrors:

There are basically three kinds of mirrors.

Plane or Concave or Converse

The arrows indicate the mirror side. The plane mirror is the one you are most familiar with. It reflects the object, like your face, but in reverse. For example, your right side is the left side in the mirror. It also reflects the proper size, and it could be stated, that the reflected image is a similar figure with the ratio of one. Another property of a plane mirror is that the reflected image is just as far behind the mirror as the object (you) is in front of the mirror. What are some of the properties of the concave and convex mirrors?

Examples of these two types of mirrors is probably in your bathroom, or kitchen in the form of a tablespoon or gravy spoon. Using one side of the spoon, as a concave mirror, and the other side, as a convex mirror write some properties of these types of mirrors. Mirrors, if available, are much better to work with.)

Comment: Local school science departments or even a drug store may have these types of mirrors. The observations will possibly be:

a. The concave will invert the image. The convex will magnify but not invert.

b. Seeing is believing?

Take a tumbler about 6-8 inches high and fill it half way with water. Look directly into the glass and place your finger on the side of the glass where the water level appears to be. Now, without moving your finger, observe from a side view the water level with position of your finger. Repeat the activity using different water levels. Write your conclusion and why?

Answer: a. The tumbler is in the shape of a cylinder and the rays are bent as they enter the water.Use the same tumbler as in part b. Fill it with water, drop a quarter in the tumbler so that it is centered. Now look directly down at the quarter.

What size does it appear to be? Larger or smaller?

Answer: A half dollar.

Ramp 13.5
The Cycloid
(Very interesting curve)

Two wheels (circles) of different size (radii) are bolted together so they are concentric. Draw a radius for the large circle from the center to the base of the large circle and mark it A, then mark where the radius crosses the small circle B. Now if the large wheel is rotated one revolution, the small wheel will also make one revolution and hence you may conclude the circumferences of all circles are the equal. But how can two different size circles have the same circumference? (Make a model of the above problem for demonstration and illustrate what actually happens when the wheel in rolled with regard to the paths for the two points.) in the small circle. This can be traced if you putt holes at the 2 points A and B and trace the two paths as the large circle is rotated.

Hint: Make a hole at A and B
Large enough to insert the tip of a pencil and trace the two curves at points A and B as the large circle completes one rotation.
Observe the two curves and you will see the solution.

Answer: The curve generated by B is a cycloid. The physical property of the cycloid is that it is the curve of quickest descent.

Research: Use the computer search machine on the Internet for information pertaining to a cycloid.

Ramp 13.6
Review Decision Making Types
(A MUST REVIEW)

"The connection between the improvement
of the human conditions and the happiness
of the human race is science and the
queen of the sciences is mathematics
NIEL POSTMAN
Technopoly-The surrender of Culture *to Technology*

A review of the major points in
EVERYDAY DECISION MAKING.

The major objective is for you to understand that all decisions are based on undefined terms, defined terms, basic assumptions and previous decisions, laws

or ordinances (theorems in geometry). The following will refresh and review the methods. These Methods for arriving at decisions range from guessing to formal logic. Everybody wants to make the correct decisions and you need to also know the possible weaknesses! Conclusions using various methods of decision-making are: Guessing

- Illusions
- Listening to experts or non-experts on all sides of issues
- Observation
- Induction
- Estimating
- Direct reasoning
- Indirect reasoning
- Statistics
- Data collection Mathematics
- Forms of an Implication
- Statements, Converses, Inverses, contrapositives which are used in ads.

The group (which you were in!) can adapt the problem to fit local experiences and make the problems more meaningful by using the multiple backgrounds, activities, interests, and future plans. The way a problem is stated can motivate others to investigate it. The reason the model of elementary Geometry was used is that it is a very simple deductive system, as the early philosophers recognized, one that all can understand!

Decisions are made by Guessing!

This is a very weak way to make a decision, it amounts to basically tossing a coin to determine the decision. The only feature of it is you have passed the decision responsibility to the some "one"

else such as a coin, but not the responsibilities resulting from the decision.

Illusions:

A method that uses what you think you see, which may not be this type is used by witnesses to an event. This plays an important role in decision making and reporting in the courts. Example: The following example is one of my favorites. Do you see a young girl or an old lady? First look from the left and then from the right.

If you were a witness, what would you report having seen?

Observation and their conclusions can be misleading.

Inductive Reasoning

This type arrives at a conclusion after investigating a few cases. This type requires record keeping to detect a pattern or trend. If the conclusion is a general one or about a future event, it may be invalid, since it is based on past events.

This is a great example!

Example 1: How many chords can be determined by N points on a circle? Draw a new circle for each case and count.

Points	Chords
2	1
3	3
What is your guess?	
4	6

Draw the case and what is your number assumption? Do you see a pattern?

6?	?
7?	?

Answer for 7 points (Predict)

8	?	?
N	?	Formula?

(Answer for 7 points is 21)

Example 2: Repeat example 1, but this time instead of comparing points and lines compare point and regions. Complete the following table.

Points	Regions
2	2
3	4
4	?
5	?

Predict and check for 4 points

6 - ?
7 - ?

Predict and check for 8 points.

Moser invented the problem around 1950.

Inductive Reasoning is drawing a conclusion as to future events from a few past events. The method for arriving at a conclusion is like predicting the future and is always questionable. The medical profession, weather forecasting, financial investors, auto repair, insurance companies, trades, many professions, and also our elected leaders use this type of reasoning to reach a conclusion.

A different type of example of an induction problem: a false conclusion! A farmer feeds his turkey every morning for a month, but on Thanksgiving Day the farmer came out to the feeding area and the turkey is in for a disappointment.

What happened?
(From a NCTM Convention.)

Many times, logical conclusions may not follow from the certain causes, since the information may be bias.

Example:

1.The junior class in a school A voted on the following question: Should cell phones be turned off in classes? (Only 15% voted no. Did this group, who voted, use phones?) Which of the following statements, printed in 5 different papers, is the most correct?

    a. High School students vote to turn off cell phones in class.85% of high school Which is students vote to prohibit cell phones in school.

    b. Most students don't use cell phones in classes

    c. 15% of students use cell phones in classes

    (Poll result needs to also report, who were polled, where, how, and when.)

Ramp 13.7
The use and misuse of implications
(A must!)

Remember: Statements are true or false
and conclusions are valid or invalid.

1. Fact: All students of school X, wear red caps at the football games. The Fact: John is at the game and is wearing a red cap. Conclusions: Which ones are valid?

    a. John must be a student of school X.

    b. Peter is not wearing a red cap at the game, then he is not a student at X.

    c. Joe is not a student at X, then he does not wear a red cap at the game.

2. Listening to opinions by experts or non-experts (TV, radio, media or people).

3. The problem with this method in #2 is that issues are restated in various ways and the terms misinterpreted as to the truth of the statements and conclusions. The advertising world and other "worlds" (such as politics) use these forms to take advantage of people who do not understand them and the implications.

A diagram to help the interpretation of A->B then B:

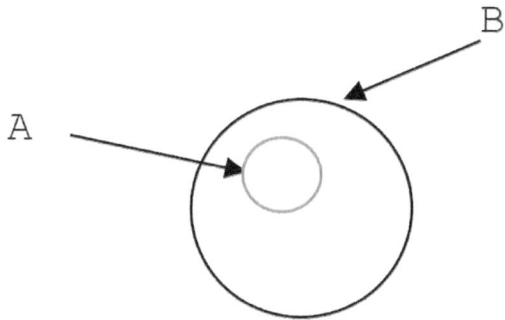

Original statement:
If A, then B.(valid)
Converse:
If B, then A. (not always valid)
Inverse:
If not A, then not B. (not always valid)
Contrapositive:
If not B, then not A. (valid)

Implications are Very important!
Be sure to understand how they are
worded and understood!

1. Fact: If you fly the Flag on Memorial Day, then you are patriotic. Which of the following are valid?
   a. If you are patriotic, then you fly the flag on Memorial Day.
   b. If you do not fly the flag on Memorial Day, then you are not patriotic.
   c. If you are not patriotic, then you do not fly the flag on Memorial Day.

Data collection

2. Polled groups are many times used to indicate a trend. Listed are questions you should know about polls. Give an example of each of the following and how it can influence the result.
   a. The number polled and reported as who, how and when.
   b. The selection process, time of day, the question, randomness, age, location)
   c. How contacted, email, telephone, U.S. mail, or interview.

3. Two witnesses can report completely different reports as to what they thought they saw. (The following was sent to the author by a friend.) Witness A reported seeing a frog. Witness B reported seeing a horse.

What do you see? (This is another very good example!)

316

The key is to understand that all conclusions are based on:

- Undefined terms
- Defined terms
- Assumptions
- Conclusions based on the above three. Valid or not!

The Declaration of Independence is a good example. (Read the first few pages and observe why independence was justified.)

Ramp 13.8
Course Review for Geometry
Be Selective

"Understanding evolves from work, appreciation is from applications."

Unknown

The geometry content will be reviewed as a separate review since the major tests like the SAT or ACT results have indicate it is the most needed.

Suggestion: short answer questions - use a ruler for drawings.

Suggestion: Let each member of your group or family justify their answers, if there are questions. The answers are given to this review.

Questions
1. Are the following statements valid definitions? Defend your answer. Is it true when reversed.)
   a. A restaurant is a place that served food.
   b. Mathematics is a useful course.

   c. Geometry is studied in all school.
   d. A plane triangle is a set of three non-collinear
      points and the line segments determined by
      the three points.
   e. If you vote, then you are well informed.
2. What are the three geometric terms in that are
   classified as undefined?
3. How many points are needed to determine a
   geometric line?
4. Draw a ray and label it AB Draw a line segment
   and label it AC.
5. Draw a triangle and label it CDE.
6. A geometric plane is determined by points.
7. What is the sum of the angles in a plane
   triangle?
8. How is the distance between points A and B
   determined?
9. What is a theorem?
10. What are the 3 conditions for 2 triangles to be
    similar?
11. When are triangles congruent?
12. a. What are parallel lines?
    b. What are skew lines? (Need a dictionary?)
13. Draw three acute scalene triangles, label each
    ABC. (Use your ruler and protractor.)
    a. In one triangle draw the three medians.
    b. In the second triangle draw the three altitudes.
    c. In the third triangle draw the three angle
       bisectors.
    d. What is conclusion?
14. The shortest distance from a point to a line is
    the distance.
15. If A implies B and A is given, then ?
16. Draw a rhombus that is not a square. (Need a
    dictionary?)
17. Draw: a. convex polygon. b. concave polygon.(Need
    a dictionary?)

18. If the sides of a triangle are 46, 23 and x, then what do you know about the measure of x?

19. The number of square units a plane figure contains is its volume?

20. The Pythagorean Right Triangle Theorem states _____.

21. What is the sum of the interior angles in each of the following figures? (Do not use a protractor. **Hint:** A triangle is 180 degrees. Do you see a relationship between the number of sides and the sum of the angles in the figures below? What is the relationship between sides and angle sum.

a.   b.   c.

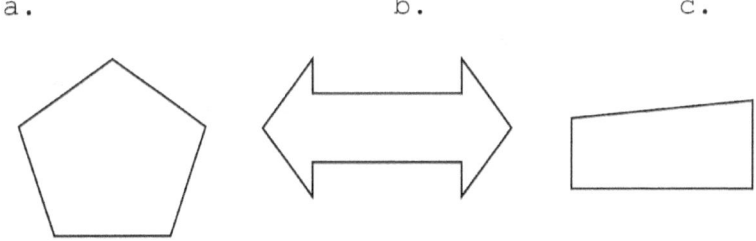

22. How many planes may four points determine?

23. The two top teams in the conference ended the season with records as shown in the following table.

Team	Losses	Wins
8	A	29
9	B	32

Team A defeated B once and Team W defeated A once. Plus team B had 3 non-conference games as indicated.

Which team would you say should receive the Cup. Why?

24. What is the perimeter of a right triangle ABC, where C is the right angle?

Given: AB = 3x units and BC = 4x units.

25. What is wrong with this argument?
A yard is 36 inches in length and ¼ yard is 9 inches. (Take the square root of is 9 inches equals is ¼ yard. Problem?

26. Students of school X voted that all students will wear a RED cap at the Saturday game. Which of the following are valid.
Hint: If-Then form.

   a. John wore a red cap, therefore he is a student of X.
   b. John is not a student of school X, therefore, he will not wear a red cap to the game.
   c. Joe did not wear a red cap to the game, therefore he is not a student of school X.

Answers

1. a. True, but not a definition.
   b. True, but not a definition.
   c. and d. Defend your answers
   e. Yes, a definition. f. Valid
2. Point, line, and plane
3. Two
4. Real number
5. Triangle CDE
6. A BC three non-co-linear points
7. 180 degrees
8. 2 points and segment. Three non- collinear points.
9. A theorem is an important mathematical statement that can be proved.
10.a. Angles are equal. (AA)
   b. The corresponding sides are proportional. (SSS)

   c. Sides proportional and included angles equal.
     (SAS)

11.Triangles are congruent if they are similar and the ratio of sides is 1.

12.a. Parallel lines are lines in the same plane and do not intersect.

   b. Skew lines are lines not in the same plane and do not intersect.

13. If each of the segments intersect in a point. Draw the figures.

14.Perpendicular distance

15.Then B

16.

 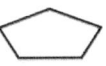

17.What is the area of each?

18.In a right triangle with sides a, b, and c, then $a^2 + b^2 = c^2$ where c is the hypotenuse.

19.a. 360 b. 720 c. 540 $S = (n-2)180$

20.4, 0

21.Four: ABC, ABD, CD, BCD,

22.Team H has a higher winning percentage, but there could other factors to be considered like Team B's and A's schedules, or the scores when they played each other.

23.12x un

24.Listen to the explanations

24.5x

25.not an equation

26."c" is valid, the contrapositive.

Suggestion: If you missed any of these problems, record the missed problem numbers, understand your errors, check the indices to refresh your memory and wait a week or two and retake the review. This review has 9 sessions and when you complete them your score will amaze you.

The proof of the pudding is in the eating. What does this statement mean?

Ramp 13.9
Conic Sections and Decision Making

Comment: You may find some of this review challenging, but isn't that the way you learn? You should take all the time you need, use your notes, calculator, and write complete Ramp answers and/or questions and review the concepts involved. (Your group may also work only a few selected problems.) Ask for explanations from others in your group as to your problems. (Look at the review again, especially a few days before your next group meeting. Remember this course content is considered the basic mathematical bridges for all professions, and more importantly, for an informed citizenship.

Problem questions should be discussed by the group. This review could take some time, depending on your recall. Group or Family discussions are very valuable with regard to how and why. Reviews are great for learning! If you have a son or daughter in high school or college they may be very help for and enjoy the discussions.

Conic Sections or the curves that control our lives.

Activity

Graph the following equations for the values indicated. Sketch the graph given the values and the and name the curve.(Use graph paper)

1. Given the equation.
   Y = 2x + 1 and the following
   Plot the points for the whole number
   values for x: −4 > +4.
   Draw the graph and name it.
1. Given equation. Y = X^2 + 2,
   Plot the r whole number values for
   x: −4 < X < 4
   Draw the graph and name the curve.

3. Given the following equation and plot values of x.y^2/3 + x^2/4 = 1: for the points −4<x<4 and draw the curve plus name the curve.

4. Given: X^2/4 + Y^2/3 = −1. Plot the points for college, integer values of x > −5 and <5., Draw the curve and name it

5. Given: x^2−y^2 =1, plot the points for the these values −4>x<4. What is the name?

Ramp 13.10
Number Sense

Group discussion and use your calculator.

1. If x is 4 and y is 3 then what is the value of:
   (one per person)2x−3y + 9x/y −5(x−4y) + (x−y)^2?

   a. x/y + y/x?
   a. (x/y)/(y/x)

   b. What is the answer to x/y, when x is divided
     by 2/3 and y is 05?

2. What is the answer to 0/5? 0
3. What is 10% of $17.50?
4. Give an example for the following:
   a. Addition of integers is commutative.
   b. Write a second-degree equation in one variable.
5. Short answer questions:
   a. The additive inverse of −5 is?
   b. The multiplicative inverse of 1/3 is?
   c. The square root of 49 is ?
   d. The $\sqrt{49}$ is ?
   e. How far is the point (5,12) from the point
     (0,0)?
   f. The tangent of 45 degrees is?
   g. The sin of 30 degrees is?
   h. If the probability of A is 3/5, then what are
     the odds in favor?
   g? If the odds are 5 to 2, then what is the
     probability?

Group Problem Solving
Be selective!

Group: <u>The answers to these problems are intentionally</u>
<u>left out</u>. Listen to the others explain their
answers, let a person justify their answer and all
will understand. This problem can be worked using
Theorem 19 on page 219.

Answers:
   See problem 1 listed the following measures for
   a triangular shaped city park: BC = 325 ft, AB
   = 210 ft, m∠ A = 75 degrees. Angle C is 38°and
   angle B is 67°. Use trig and your calculator.
   Angle A is 75°
   Side c is 210feet.

Side A is 325 feet and using Theorem 19 solves angle C and is 38°.
We need to know the measure of side b. Angle C is 67° by Theorem 19
Now draw a more accurate shape of the lot.

Questions:
   a. AC is between what two measures or values?
   b. What is the length of AC?
   c. What is the measure of angle C?

Answer: 38°

2. What is the area of the triangle?
                  Answer: 21210 sq. ft (rounded)

3. If a bag of fertilizer sells for $10 and covers 8000 square feet, then what is the cost to fertilize the park area once?
                  Answer: $30

3. The term acre-foot is used for measuring the amount of water in flooded areas. It means an acre of water one foot deep. How many gallons of water are in an acre-foot? (A cubic foot of water contains 7.4805 gallons and an acre foot is 43560 gallons.)

5. The perimeter P of a square varies directly as the side S.
   a. Show a few cases to illustrate this.
   b. Write the equation for the relationship between P and S.

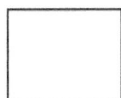

6. The load that a 2x4 with support varies inversely to the length between the supports. Example: A one

foot 2x4 will support 1280 Lbs. Where a 2 foot 2x4 will support 640 Lbs.
  a. Write the equation.
  b. What may an eight foot 2x4 support?
  c. Does it make any difference as to which side, the 2 or the 4, is turned down? (Justify your answer using a yardstick.)

7. Your uncle gives you $5000 when you are 6 years old and deposits it in bonds which pay 8.5% interest compounded annually. What will be the value, when you are 18? How many years will it take for the money to double? (Banker's Rule of 72.) $10645.48)

8. Your aunt agrees to deposit in an $200 per month at 2% when you turn 10 to help with college expenses. She will do this for eight years for 72 months at 2% per month. (72 months interest rate of 2%. What will be the value of the annuity at the end of the eighth year? Justify your answer

9. Name the four curves identified as the conic section and draw a sketch each.

10. If the mean for two classes (A and B) are the same, but the median for class B is higher, then which class do you think is better using the same test and teacher.)

11. If the means for two classes (C and D) are the same, but the standard deviation for class C is smaller. Then which class do you think is better and why? (Same subject, same teacher and same test.)

12. Would you want son's teacher to grade on the curve?" Explain your answer.

13. What are the 2 interpretations of "or" in A or B.

14. Interpretation depends on definition of words. Which words need defining in this quote? "I can assure you that my country will never take any action unless there is a serious provocation by your country". What word needs defining?

Check answers and have their solutions explained.

Ramp 13.11
Final Critical Thinking Review

This will be great for a small group to discus and solve for the answers. People (parents) should not tell others (kids) what to think, but help them to learn how to arrive at a conclusion from facts.

Decision Making Review

As mentioned before the teaching of Mathematics has two major objectives:

Teaching the mathematical skills needed for whatever the professional background calls for. This means being a student is a life time obligation. The skill of Decision Making for a better life. This was stated at the entrance to Plato's Academy and the author changed it to the following!

LET NO PERSON EXIT HERE IGNORANT OF BASIC PLANE AND SOLID GEOMETRY TAUGHT FOR DECISION THINKING!

To meet the needs of modern day requirements for better decision making this has been changed to:

**LET NO PERSON EXIT HERE IGNORANT
OF BASIC DECISION MAKING.**

An example of the use and problems with polls was recently on TV.

The following was stated with regard to the issue of Gun control.

Would you favor background checks at?
            92% at the store
            87% at gun show
            75% private sellers

Would you favor the following for gun safety? (Yes) 69% gun registration, 58% ban on clips, 56% ban on assault guns, 52% ban on Ammo sales.

Would you favor or oppose armed guards in schools?
            54% favor   45% oppose

Do you own a gun? 49% Yes 49% no

Who is to blame for gun violence in the USA?(A poll consisting of the following.)
            37% parents
            37% pop culture
            23% availability of guns

This is only a part of the survey, but what is very important in evaluating the results of a survey was stated in very fine prints and hard to read is the following. The survey was listed as international, conducted by phone on Jan. 14 & 15 (Sunday and Monday) of 814 adult Americans selected at random.

Would you draw an international general conclusion base on 814 adults? What would you like to know about these adults?

  1. Were they hunters?

2. Where they live?
3. Age
4. Profession or career.
5. Their education show?

Discuss the responses to the above!

Ramp 13.12
More use and misuse of Implications

1. From the following statement write two conclusions that are valid and two that are invalid.

The State Good Driver Association indicated that very few drivers were killed at speeds over 100 mph.

Letters to the editor stated the following conclusion.

a. (your valid conclusion)
b. (your valid conclusion)
c. (your invalid conclusion)
d. (your invalid conclusion)

2. Three best student had a very good records and the class ask the teacher which one she thought is the most intelligent decision maker? The teacher set up the following test. She instructed each of the three to stand in a corner of the room. She then blind folded each of them and carefully explained she would put a black or white hat on each of them. When the blind folds are removed, they are to put their right arm above their head if they see a black hat and they were to take their hand down when they know the color of their hat. A black hat was put on each and when they took off the blind folds they all to each put their right hand above their head. After a short while one student took his hand down

and said, "I have a black hat on!" How did he or she know?

Suggestion: Have three black hats(or any color) and act this out in your group for a complete and easy understanding!

3. All conclusions are based on 4 terms. What are the four terms?

4. Golden Rectangle (optional, but used in the ART Profession) and by most people unknowingly also use your computer search.

The rectangle is classified as a Golden Rectangle, the following equation is valid L/W = W/L-W.

What is the ratio of the sides? Hint: Assume the length 1 and then is .618 the w side measures one unit, then you have an equation in one variable and solving the length is 1.618 and the width is .618.

Draw a golden rectangle using the L as 1 foot and the width as .6. of a foot.

Suggestion: Talk to an Art teacher about the Golden Rectangle.

Bridge 14: The Last Word

This is the end of MATH BRIDGES TO A BETTER FUTURE. The author sincerely hopes you feel it was worth the experience, especially the conversations for the family relations and the after dinner discussions (about one hour per week with the family).

The family found it very informative, especially the backgrounds based on mathematics about 2500 years ago.

Remember all conclusions are based on, undefined terms, defined terms, Postulates and Theorems or Laws based on the first three.

CONGRATULATIONS and here is to a BETTER FUTURE in coming Years and helping others to use Bridges and information leading to the Bridges called Ramps I mentioned in the prologue that the four past presidents (Johnson, Bush #1, Clinton, and Bush #2) and their Foundations for their history to establish a leadership school at, I think, the University of Texas, which was to have the program to educate a select group each year (like a graduate program to possible produce more qualified leadership.) Each year since 2015, about 100 have completed the program. There has been about 800 applicants also each year. Unfortunately, there has been no information as to the program except the graduation honor program each year. Wouldn't you like to hear about the program? I would! These four past presidents were so concerned that their Foundations supplied the funds to create and support this unique program which hopefully will result in better potential leaders that are better informed persons in the government, especially in the elected positions.

I sincerely hope the weekly discussions have worked to benefit your family and especially the children. I suggest you give the book to your youngest brother or sister to carry on the relationship. From my experience the young people were most appreciative and the older ones continued their correspondence over the years with pictures and the habit will continue with future generations.

Jim Elander

Appendix 1: Definitions

Definition A: A Theorem is an important mathematical statement that follows logically from a set of undefined terms, definitions, postulates or other theorems.

Definition B: Postulate is a statement, which is assumed to be true.

Definition 1: The set of real numbers is the union of teratomas and the irrationals.

Definition 2: A Rational number is a real number that can as the ratio of two integers.

Definition 3: An irrational number is a real number, which cannot be written as a ratio of two integers.

Definition 4: Is 0 rational, irrational, or even a number?

Definition 5: Inductive reasoning is arriving at a conclusion you think is true after observing a few cases.

Definition 6: Arithmetic Sequence: A set of terms such that given the first term, the next one is obtained by adding a constant term. N+1 = N + K

Definition 7: Arithmetic Series: An arithmetic series is the indicated sum of an Arithmetic Sequence.

Definition 8: A geometric sequence is a set of terms such that each term is equal to the preceding term multiplied by a constant factor. N+1 = K(n)

Definition 9: A geometric series is the sum of a geometric sequence.

Definition 10: Simple interest is the money paid for the use of the money borrowed. The formula for simple interest is: $I = prt$.

Definition 11: Compound interest is interest not only paid on the loan but on the interest also. Formula for the total amount is $A = p(1 + r/n)^{ny}$.

Definition 12: An ANNUITY is series of equal payments or deposits made for an agreed amount of time at a specified rate of interest.

Definition 13: A permutation is the way a set of objects can be arranged where order does matter. (Example: AB and BA are two permutations.)

Definition 14: N! read N factorial, means to multiply all the positive integers up to and including the number N.
$$N! = 1x2x3x4x5x \ldots x(N-2)(N-1)N$$

Definition 15: A COMBINATION is the way a set of objects can be arranged where order does NOT count. Example: AB and BA are one combination.)

Definition 16: An equation is a statement that two numbers are equal.

Definition 16a: Probability of an outcome is the ratio of the number of favorable outcomes or successes to (or divided by) the total number of possible Outcomes.

Definition 16b: Empirical probability is the probability based on actual trials.

Definition 16c: The EXPECTATION of an event (e) is the probability of the event P(e) times the number of trials (T). Formula: E(e) = [P(n)](T).

Definition 16d: A game is fair when the cost of playing equals the expected probability value times the winnings, or n(c$) = n(w$)[P(w)], where: c$ and w$ is the cost per play or win. n is the number Example: It cost $2 to toss a coin and if a head turns up you win $4. If you play 4 times, is it a fair game.

Definition 16e: The formula for odds in favor of an event e is: Odds = P(e)/[1-P(e)].

Definition 17: An equation is a statement indicating two numbers are equal.

Definition 18: A formula is an equation representing a general rule such as a theorem.

Definition 20: A line is defined as the set of points that satisfy the equation y = mx + b, where x, y, and b are real numbers with the restriction that x and y both cannot be zero.

Definition 21: Direct Variation. If two variables are so related that y = kx, then x and y vary directly where k is the constant that relates the two variables. (This also indicates y/x = k.)

Definition 22: The SINE (SIN) of an acute angle in a right triangle is the ratio of the length of the side opposite the angle divided by the length of the hypotenuse.

Definition 23: The COSINE of an acute angle in a right triangle is the ratio of the length of the adjacent side divided by the length of the hypotenuse.

Definition 24: The Tangent of an acute angle in a right triangle is the ratio of the length of the side opposite the angle divided by the length of the side adjacent to the angle.

Definition 25: Trig functions defined for the general angle. (The vertex of the angle is at the origin, one ray on the positive x-axis and the other ray is rotated counterclockwise. The "distance" in the following definitions refers to distance the point is from the origin.

    a. Sin A = y/R (opposite/hypotenuse or ordinate/ distance)
    b. Cos A = x/R (adjacent/hypotenuse or abscissa/ distance)
    c. Tan A = y/x (opposite/adjacent or ordinate/ abscissa)

Definition 26: An equation is an identity if the equation is true regardless of the value that is substituted for x or the variable.

Definition 36: A Quadratic equation in two variables is defined as y = = Ax^2 + Bx + C, where A, B and C are rational numbers and A is not equal to 0.

Definition 37: An equation is a function if its set of ordered pairs (x,y) satisfies the condition that for each x value there is only one y value.

Definition 38: A parabola is the set of all points in a plane that are equal distance from a point (focus point) and a line called the directrix.

Definition 39: The $\sqrt{-1}$ times the $\sqrt{-1}$ equals -1.

Definition 40: The symbol for the $\sqrt{-1}$ is i.

Definition 41: A + Bi is a complex number; A and B are real numbers and i is $\sqrt{-1}$.

Definition 42: A circle is the set of points on a plane, which are equal distance from a given point called the center.

Definition 43: The locus or geometric definition of an ellipse is the set of all points P on a plane such that given two points F_1 and F_2, then the set of points P that meet the condition $PF_1 + PF_2 = K$ (K is a positive constant number.) is an ellipse.

Definition 44: Given two points (focus points or foci) the set of points which meet the following condition is a hyperbola, $F_1P-F_2P = K$. This means the distance any point P is from one focus point minus the distance the point P is from the other focus point is a constant.

Definition 44.a: A square hyperbola is one where the basic graphing rectangle related to the hyperbola is a square.

Definition 45: Inverse Variation results when two variables are related in such a way that their product is always the same number or constant. Symbolically, this is stated as xy = k.

Definition 30: The average speed is the total distance divided by the time it takes to travel the distance. S or R (rate) = D/T

Definition 31: The MEAN or arithmetic average for a set of numbers is the sum of the numbers divided by n, the number of numbers in the set.
Formula: Mean = (sum of n scores)/n

Definition 32: The MODE for a set of data is the most popular or most frequently occurring element in the set.

Definition 33: The MEDIAN for a set of data is the middle element when the elements are arranged in order of size or magnitude.

Properties of the Normal Curve

1. The mean, mode, and median are all equal or the same value and occur at the center or the line of symmetry.

2. One STANDARD DEVIATION on each side of the mean (line of symmetry) will include approximately 68% of the data.

3. Two STANDARD DEVIATIONS on each side of the mean (line of symmetry) will include approximately 95% of the data.

4. Three STANDARD DEVIATIONS on each side of the mean (line of symmetry) will include approximately 99.8% of the data.

5. From the data or the curve, the RANGE can also be determined.

Definition 34: The RANGE is the difference between the highest or largest number and the lowest or smallest number in the set of data.

Definition 35: STANDARD DEVIATION is the square root of the mean of the squares of the deviations from the mean.

Definition 36: Properties of exponents

a. If x^m times x^n, where x is real and n and m are rational, then the product is x^{m+n}.
b. If x^m/x^n, where x is real and not equal to zero, and n and m are rational, then the quotient is x^{m-n}.
c. x^0 is equal to 1 providing x is not zero.
d. If x^{-n}, where x is real and not 0, and n is rational, then x^{-n} equals $1/x^n$.
e. If $x^{n/d}$ and x is real, and n and d are rational and d is not zero, then the expression equal to the square root of n/d.
f. If $(x^m)^n$ where m and n are rational numbers, then the product is x^{mn}.

Definition 49: If $n = b^L$, then the logarithm of n to the base b is L, and conversely. ($\log_b n = L$)

Definition 49a: If Log n with no base indicated, then it means the base is 10.

Definition 49b: Products by logs Log of (A times B) is the $\log_b A$ + the $\log_b B$. (The bases must be identical.)

Definition 49c: Quotients by logs Log of (A divided by B) is the $\log_b A$ - the $\log_b B$.

Definition 49d: Powers by logs Log of A^n is n times $\log_b A$.

Appendix 2: Postulates

Postulate 1: Two points will determine one and only one straight line.

Postulate 2: Three non-collinear will 3 determine a geometric plane.

Postulate 3: In a triangle the sum of 2 sides is greater than the third side.

Postulate 4: The area of a rectangle is area of base times height.

Postulate 5: There is a one - to one correspondence or matching between the points on a line and the real numbers.

Operation Postulates 6: Add or subtract any two numbers and you get an answer.
Multiply or divide (except division by 0) any two numbers and you get an answer.

Commutative Postulate 7: For any two numbers A + B equals B + A.(Addition is commutative.)

Commutative Postulate 8: For any two numbers A x B equals B x A.
   (Multiplication is commutative)
  Mathematicians name 3 & 4 the
     Commutative Postulates.)

Addition Postulate 9: For any two numbers A + (-A) = 0

Postulate 10: For any number A x (1/A) = 1 providing A is not equal to zero.

Postulate 11: For any number A + 0 = A.

Identity Postulate 12: For any numbers A x 1 = A. A is not 0.

Associative Postulate 13: For any three numbers A + (B + C) = (A + B) + C.

Associative Postulate 14: For any three numbers (A x B) x C = A x (B x C).

Distributive Postulate 15: For any three numbers A(B + C) = AB + AC.

Postulate 16: The area of a rectangle is the LENGTH times the WIDTH.

Postulate 17: The volume of cube is the area of the base times height.

Fundamental Counting Postulate 18: The number of permutations of n different objects taken all at a time is n!

Postulate 19: The formula for the number of permutations of n different objects taken r at a time is P(n,r) is = n!/(n-r)!

Postulate 20: The formula for the number of permutations of n objects taken all at one time where r objects are the same, s are the same, and t are the same, etc, is n!/(r!s!t!).

Postulate 21: The number of combinations of n different objects taken at a time is:

$$C(n,r) = \frac{n!}{r!(n-r)!}$$

Postulate 22: If the statements for successful cases are joined by OR, then the probabilities for each case are added.

Postulate 23: If the statements for the successful cases are joined by AND, then the probabilities for each case are multiplied.

Postulate 24: If given an equation, then you can add a number to each side of the equation.

Postulate 25: If given an equation, then you can multiply each side by the same non- zero number. (This includes division, powers and roots.)

Postulate 26: If given a set of equations, then the equations can be added (subtracted) and the result is an equation.

Postulate 27: If given an inequality, then a number can be added (subtracted) to each side of the inequality and the result is an inequality of the same order.

Postulate 28: If given an inequality, then the inequality can be multiplied (divided) by a positive number and the result is an inequality of the same order.

Postulate 29: If given a set of inequalities of the same order, then the inequalities can be added and the result is an inequality of the same order.

Appendix 3: Theorems in BRIDGES

Theorem 1: Three non collinear points Will determine a geometric plane.

Theorem 2: Given a triangle with sides a,b,and c, then $a + b > c$, and $a > c-b$ and $a<c+b$.

Theorem 3: Pythagorean theorem $a^2 + b^2 = c^2$

Theorem 4: The area of a parallelogram is B times H.

Theorem 5: The area of a triangle is ½ the base times the altitude.

Theorem 6: The sum of an arithmetical series is:
   $S = (n/2)(a + L)$ or$[(n/2)(2a + (n-1)d]$
where: "a" is the first term "n" is the number of terms "d" is the constant difference "L" is the last term.
$$L = a +(n-d).$$

Theorem 7: The volume of pyramid is 1/3 times area of base time the height.

Theorem 8: The sum of a geometric series is $S = a(1 - r^n)/(1-r)$ or $S =(a - ar^n)/(1 - r)$ where "a" is the first term, "r" is the ratio or common factor.

Theorem 9: The sum of an infinite geometric series is $S = a/(1-r)$, where r is between 0 and 1 $(0 < r <1)$.

Theorem 10: Compound Interest formula is $T = A(1+R)$ $n^{x\theta}$ where x is the rate of percent and θ is the term in years.

Theorem 11: The Annuity formula:

A = np[(1+r/n)$^{ny+1}$ - (1+r/n)]/r where: A is the value
at the end of the contract term.
     p is the payment made each period.
     r is the annual rate of interest.
     n is number of periods per year.
   y is the number years

Theorem 12: If given an inequality, then the
inequality can be (divided) by a negative number
and the result is an inequality of the opposite
order.

Theorem 13: A proof for Pi = circumference divided
by diameter

Sine Theorem 14: If the sides of a triangle ABC are
a, b, and c, then Sin A/a = sinB/b = sinC/c.

Cosine Theorem 15: If the sides of a triangle are
a, b, and c, then:
   $a^2 = b^2 + c^2 - 2bc(Cos A)$ or
   $b^2 = a^2 + c^2 - 2ac(CosB)$ or
   $c^2 = a^2 + b^2 - 2ab(CosC)$

Theorem 16: The volume of a cone is 1/3) area of
Base times altitude.

Theorem 17: The pole theorem (p. 219)

Theorem 18: The Jordan Theorem (p. 220)

Theorem 19: Given any angle, then $Sin^2A + Cos^2A = 1$
(Trig form of Pythagorean Theorem)

Theorem 20: If A is an angle then Tan A = Sin A/
Cos A for values of A except for odd multiples of
90 degrees.

Theorem 21: Sin A/a = Sin B/b = Sin C/c

Theorem 22: If given the quadratic equation, Ax^2 + Bx + C = 0, then the solution is:

$$x = \frac{-B \pm \sqrt{(B^2 - 4AC)}}{2A}$$

Theorem 23: Given the quadratic equation, $y = Ax^2$ +Bx + C, then the turning point on the graph has the coordinates x=(-B/2A,y= (-B/2A)). The equation for the line of symmetry is x = -/2A.

A few theorems for the conic curves

Theorem 24: The formula for a circle is: {(x,y): x^2 + y^2 = r^2, where r is the radius and the center is at the origin.}

Theorem 25: The equation for an ellipse is x^2/a^2 + y^2/b^2 = 1, where the center is at the origin and the foci are at $\pm \sqrt{(a^2-b^2)}$ on the major axis.

Theorem 26: The equation for a hyperbola is:
$$x^2/a^2 - y^2/b^2 = 1$$

Focus points are:
$$((a^2+ b^2), \ 0),(- \ (a^2+ b^2),0)$$
The x-intercepts are at (-a,0),(a,0).
The basic rectangle (for sketching purposes) is determined by the points: (a,b), (a,-b), (-a,-b) and (-a,b).

Appendix 4: Conversion Tables

Measures of Length

English		Metric
1 inch (in)	=	2.54 centimeters (cm)
0.39 inches	=	1 cm = 0.1 decimeter (dm)
12 in, 1 foot (ft)	=	30.48 cm
3 ft = 1 yard (yd)	=	0.9144 cm
39.37 in	=	1 meter (m) or 100 (cm)
5280 ft = 1 mile	=	1.609 kilometers (km)
0.621 mile	=	1 km
1 nautical mile	=	1.1508 miles = 1.852 km

Measures of Area

English	Metric
1 sq ft= 144 sq in = 929.03 sq.cm =.092903 sq m	
1 sq yard = 9 sq ft = 0.836 sq. meters (m)	
1.196 sq yd = 10.765 sq ft = 1 sq meter	
1 sq mile = 640 acres	= 259 hectares
2.471 acres	= 1 hectare
1 acre = 43,560 sq. ft	= 0.405 hectare

Measures of Volume

English	Metric
1 cubic in	= 16.388 cubic cm
1 cubic ft = 1,728 cubic in	= 28,318.46 cubic cm
1 cubic yd = 27 cubic ft	= 0.765 cubic meters
1.308 cubic yds = 35.31 cu ft = 1 cubic meters	
1 U.S. fluid Gal = 4qts = 8 pts = 3.785 liters	
1 British gal = 1.06 U.S. gal	= 4 liters

(This is why a gallon of gas in Canada costs more than a U.S. gallon.)

A few kitchen conversions

English		Metric
1 teaspoon	=	5 milliliters (ml)
1 tablespoon	=	15 ml
1 ounce (oz)	=	29.6 ml

Appendix 5: For Your Own Family

Photos and family history

Put your own photos here with dates where, names, and any other information that you feel is important to remember years from now!

The End

www.ingramcontent.com/pod-product-compliance
Lightning Source LLC
Chambersburg PA
CBHW021348210526
45463CB00001B/27